Microsoft Azure

Build, manage, and scale cloud applications using the Azure Infrastructure

Mikey Lindsey

Table of Contents

Section I: Microsoft Azure

Microsoft Azure is a cloud computing platform that provides a wide variety of services that we can use without purchasing and arranging our hardware. It enables the fast development of solutions and provides the resources to complete tasks that may not be achievable in an on-premises environment. Azure Services like compute, storage, network, and application services allow us to put our effort into building great solutions without worrying about the assembly of physical infrastructure.

This book covers the fundamentals of Azure, which will provide us the idea about all the Azure key services that we are most likely required to know to start developing solutions. After completing this book, we can crack job interviews or able to get different Microsoft Azure certifications.

What is Azure

Microsoft Azure is a growing set of cloud computing services created by Microsoft that hosts your existing applications, streamline the development of a new application, and also enhances our on-premises applications. It helps the organizations in building, testing, deploying, and managing applications and services through Microsoft-managed data centers.

Azure Services

- ❖ **Compute services:** It includes the Microsoft Azure Cloud Services, Azure Virtual Machines, Azure Website, and Azure Mobile Services, which processes the data on the cloud with the help of powerful processors.
- ❖ **Data services:** This service is used to store data over the cloud that can be scaled according to the requirements. It includes Microsoft Azure Storage (Blob, Queue Table, and Azure File services), Azure SQL Database, and the Redis Cache.
- ❖ **Application services:** It includes services, which help us to build and operate our application, like the Azure Active Directory, Service Bus for connecting distributed systems, HDInsight for processing big data, the Azure Scheduler, and the Azure Media Services.

- ❖ **Network services:** It helps you to connect with the cloud and on-premises infrastructure, which includes Virtual Networks, Azure Content Delivery Network, and the Azure Traffic Manager.

How Azure works

It is essential to understand the internal workings of Azure so that we can design our applications on Azure effectively with high availability, data residency, resilience, etc.

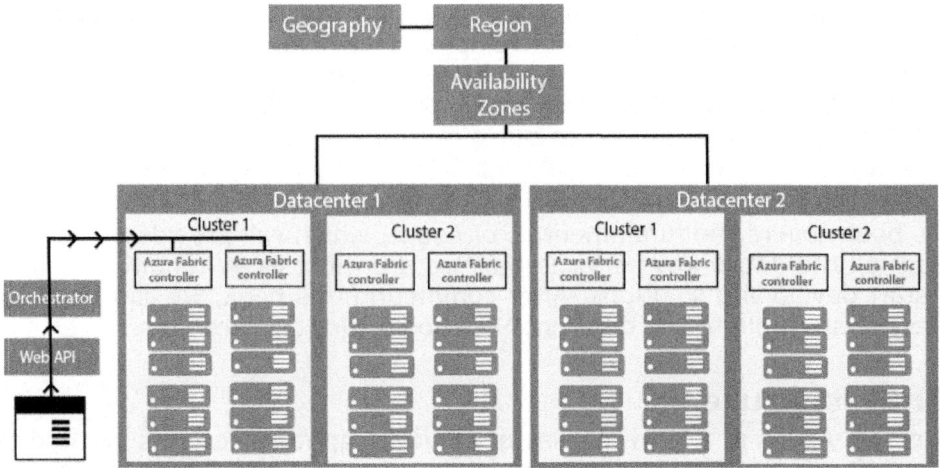

Microsoft Azure is completely based on the concept of virtualization. So, similar to other virtualized data center, it also contains racks. Each rack has a separate power unit and network switch, and also each rack is integrated with a software called Fabric-Controller. This Fabric-controller is a distributed application, which is responsible for managing and monitoring servers within the rack. In case of any server failure, the Fabric-controller recognizes it and recovers it. And Each of these Fabric-Controller is, in turn, connected to a piece of software called Orchestrator. This Orchestrator includes web-services, Rest API to create, update, and delete resources.

When a request is made by the user either using PowerShell or Azure portal. First, it will go to the Orchestrator, where it will fundamentally do three things:

1. Authenticate the User
2. It will Authorize the user, i.e., it will check whether the user is allowed to do the requested task.
3. It will look into the database for the availability of space based on the resources and pass the request to an appropriate Azure Fabric controller to execute the request.

Combinations of racks form a cluster. We have multiple clusters within a data center, and we can have multiple Data Centers within an Availability zone, multiple Availability zones within a Region, and multiple Regions within a Geography.

- ❖ **Geographies:** It is a discrete market, typically contains two or more regions, that preserves data residency and compliance boundaries.
- ❖ **Azure regions:** A region is a collection of data centers deployed within a defined perimeter and interconnected through a dedicated regional low-latency network.

Azure covers more global regions than any other cloud provider, which offers the scalability needed to bring applications and users closer around the world. It is globally available in 50 regions around the world. Due to its availability over many regions, it helps in preserving data residency and offers comprehensive compliance and flexible options to the customers.

Availability Zones: These are the physically separated location within an Azure region. Each one of them is made up of one or more data centers, independent configuration.

Azure Pricing
It is one of the main reasons to learn Microsoft Azure. Because Microsoft is providing free Credits in the Azure account to access Azure services for free for a short duration. This credit is sufficient for people who are new at Microsoft Azure and want to use the services.

Microsoft offers the **pay-as-you-go** approach that helps organizations to serve their needs. Typically the cloud services will be charged based on the usage. The flexible pricing option helps in up-scaling and down-scaling the architecture as per our requirements.

Azure Certification

Microsoft Azure helps to fill the gap between the industry requirement and the resource available. Microsoft provides Azure Certification into three major categories, which are:

Azure Administrator: Those who implement, monitor, and maintain Microsoft Azure solutions, including major services.

Azure Developer: Those who design, build, test, and maintain cloud solutions, such as applications and services, partnering with cloud solution architects, cloud DBAs, cloud administrators, and clients to implement these solutions.

Azure Solution Architect: Those who have expertise in compute, network, storage, and security so that they can design the solutions that run on Azure.

All these certifications are divided into different levels. If anyone is planning to get certified, then he/she first has to get an associate-level certification and then go for the advanced level.

Prerequisite

Before Learning AWS, one should have basic knowledge of cloud computing and computer fundamentals.

Audience

Our Microsoft Azure book is designed for students and working IT professionals who are new to Cloud Computing and want to pursue or switch their career path as Microsoft Azure Developer or Administrator.

Scope of this book

We will see the overview of cloud computing, the inner working of Azure, and how azure allocate resources. After that, we will dive into the different areas of Azure services i.e., Storage services, Compute services, Network services, App services, Data Bases, Analytics, Integration services, IoT, Security services, Monitoring and Diagnostics, and Tools. This book also provides the idea about creating VMs, website and storage accounts, etc.

Introduction to Cloud Computing

Cloud Computing is the delivery of computing services such as servers, storage, databases, networking, software, analytics, intelligence, and more, over the Cloud (Internet).

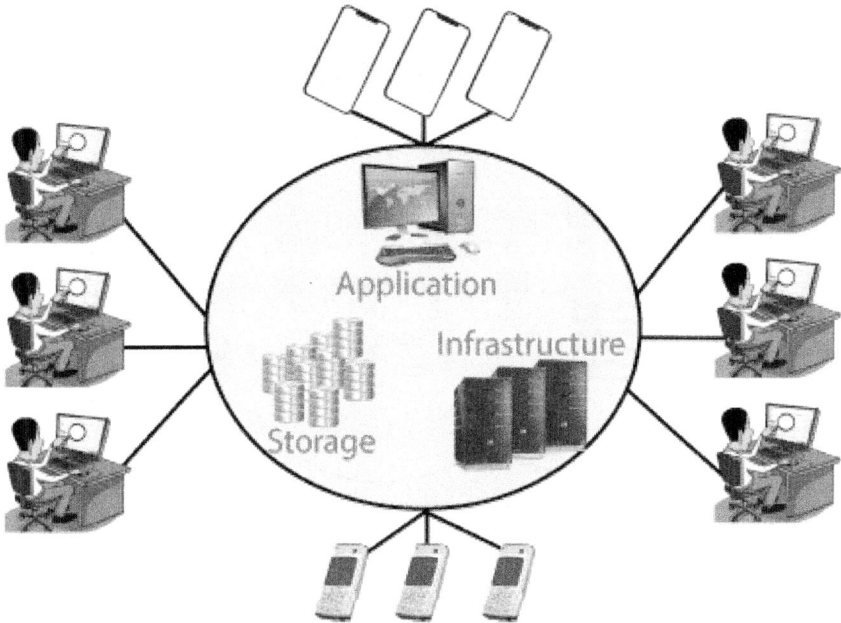

Cloud Computing provides an alternative to the on-premises datacentre. With an on-premises datacentre, we have to manage everything, such as purchasing and installing hardware, virtualization, installing the operating system, and any other required applications, setting up the network, configuring the firewall, and setting up storage for data. After doing all the set-up, we become responsible for maintaining it through its entire lifecycle.

But if we choose Cloud Computing, a cloud vendor is responsible for the hardware purchase and maintenance. They also provide a wide variety of software and platform as a service. We can take any required services on rent. The cloud computing services will be charged based on usage.

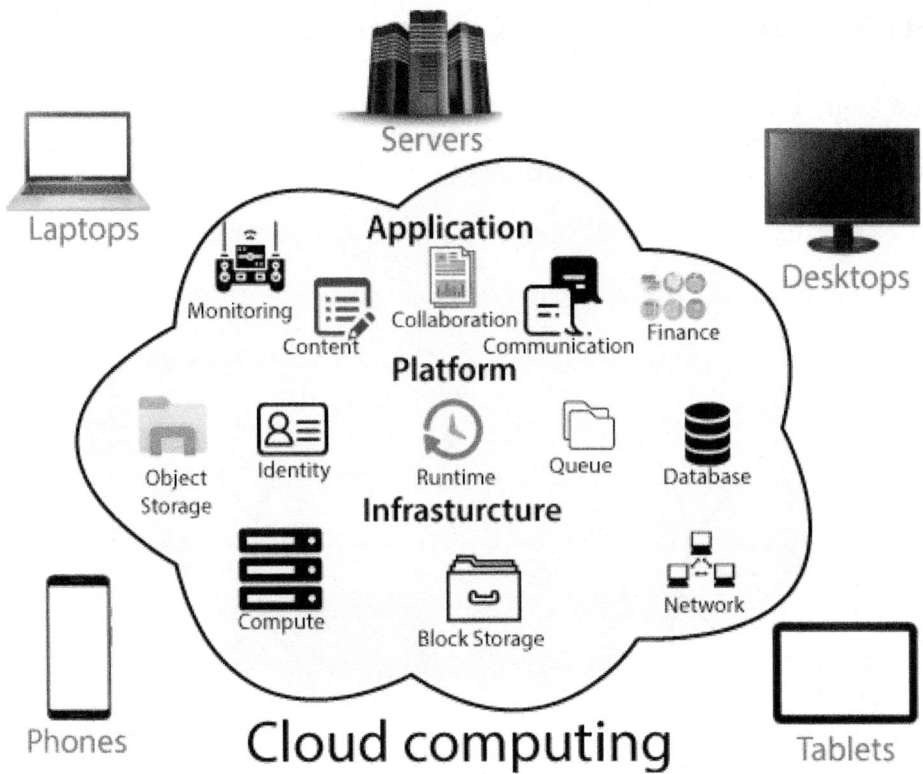

Cloud computing

The cloud environment provides an easily accessible online portal that makes handy for the user to manage the compute, storage, network, and application resources. Some cloud service providers are in the following figure.

Advantages of cloud computing

❖ **Cost:** It reduces the huge capital costs of buying hardware and software.

❖ **Speed:** Resources can be accessed in minutes, typically within a few clicks.

❖ **Scalability:** We can increase or decrease the requirement of resources according to the business requirements.

❖ **Productivity:** While using cloud computing, we put less operational effort. We do not need to apply patching, as well as no need to maintain hardware and software. So, in this way, the IT team can be more productive and focus on achieving business goals.

❖ **Reliability:** Backup and recovery of data are less expensive and very fast for business continuity.

❖ **Security:** Many cloud vendors offer a broad set of policies, technologies, and controls that strengthen our data security.

Types of Cloud Computing

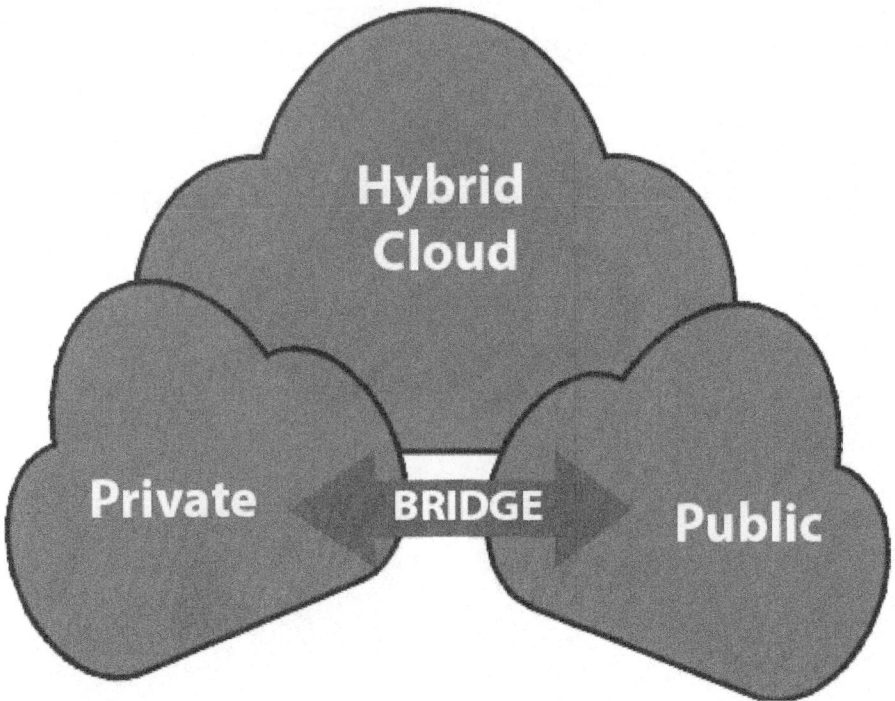

❖ **Public Cloud:** The cloud resources that are owned and operated by a third-party cloud service provider are termed as public clouds. It delivers computing resources such as servers, software, and storage over the internet

❖ **Private Cloud:** The cloud computing resources that are exclusively used inside a single business or organization are termed as a private cloud. A private cloud may physically be located on the company's on-site datacentre or hosted by a third-party service provider.

❖ **Hybrid Cloud:** It is the combination of public and private clouds, which is bounded together by technology that allows data applications to be shared between them. Hybrid cloud provides flexibility and more deployment options to the business.

Types of Cloud Services

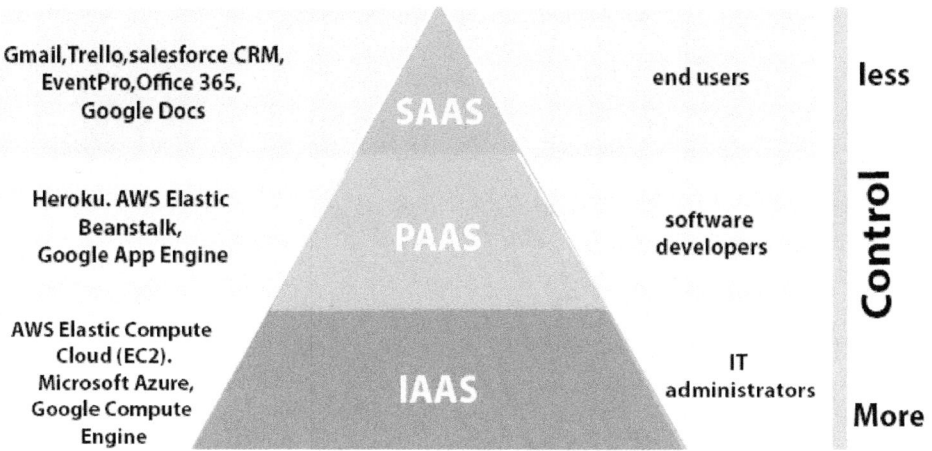

Gmail, Trello, salesforce CRM,
EventPro, Office 365,
Google Docs

SAAS

end users

less

Heroku. AWS Elastic
Beanstalk,
Google App Engine

PAAS

software
developers

Control

AWS Elastic Compute
Cloud (EC2).
Microsoft Azure,
Google Compute
Engine

IAAS

IT
administrators

More

❖ **Infrastructure as a Service (IaaS):** In IaaS, we can rent IT infrastructures like servers and virtual machines (VMs), storage, networks, operating systems from a cloud service vendor. We can create VM running Windows or Linux and install anything we want on it. Using IaaS, we don't need to care about the hardware or virtualization software, but other than that, we do have to manage everything else. Using IaaS, we get maximum flexibility, but still, we need to put more effort into maintenance.

❖ **Platform as a Service (PaaS):** This service provides an on-demand environment for developing, testing, delivering, and managing software applications. The developer is responsible for the application, and the PaaS vendor provides the ability to deploy and run it. Using PaaS, the flexibility gets reduce, but the management of the environment is taken care of by the cloud vendors.

❖ **Software as a Service (SaaS):** It provides a centrally hosted and managed software services to the end-users. It delivers software over the internet, on-demand, and typically on a subscription basis. E.g., Microsoft One Drive, Dropbox, WordPress, Office 365, and Amazon Kindle. SaaS is used to minimize the operational cost to the maximum extent.

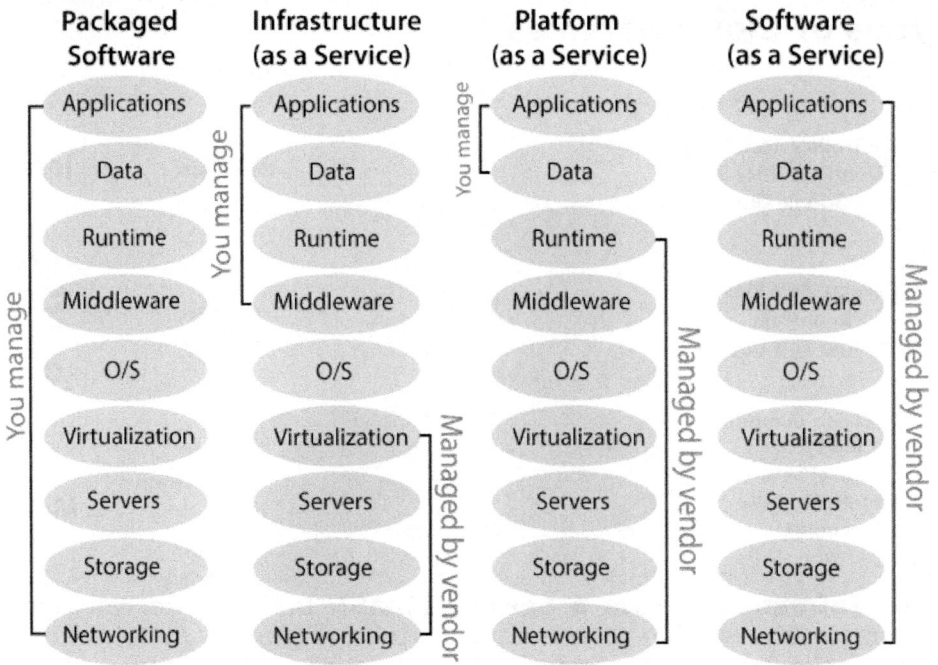

Azure Portal Overview

Azure portal is a platform where we can access and manage all our applications at one place. We can build, manage, and monitor everything from simple web-apps to complex cloud applications using a single console.

So, first of all, to log into the Azure portal, we need to register. And, if we are registering for the first time, we will get 12 months of popular free services. And also, depending on the country, we will get some amount of free credit that needs to be consumed within 30 days. And in addition to all these things, we will get some services that are free forever.

So, make sure you are completely ready to try all the services before you register for Azure because that credit is only available for 30 days.

Creating an Azure Account

Step 1: Open https://azure.microsoft.com/en-us/free/ then click on Start free; it will redirect you to the next step.

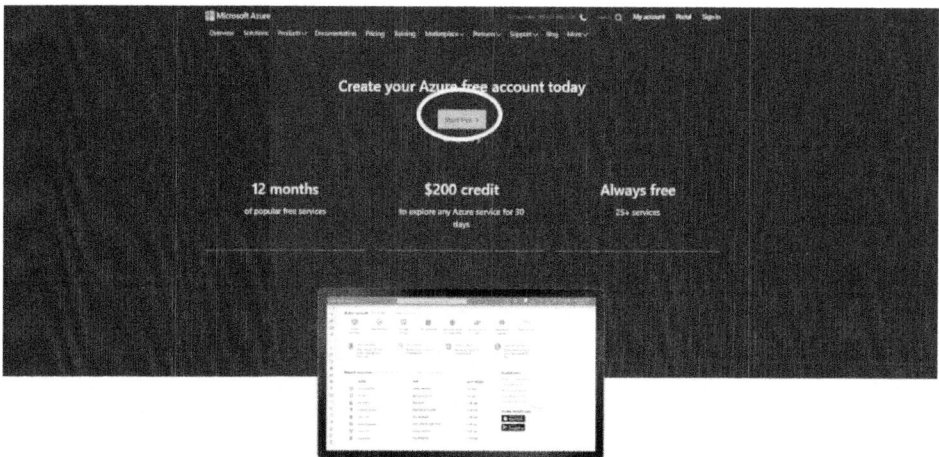

Step 2: It will ask you to login with your Microsoft account. If you already have a Microsoft account, you can fill the details and login. And if you don?t have one, you must signup first to proceed further.

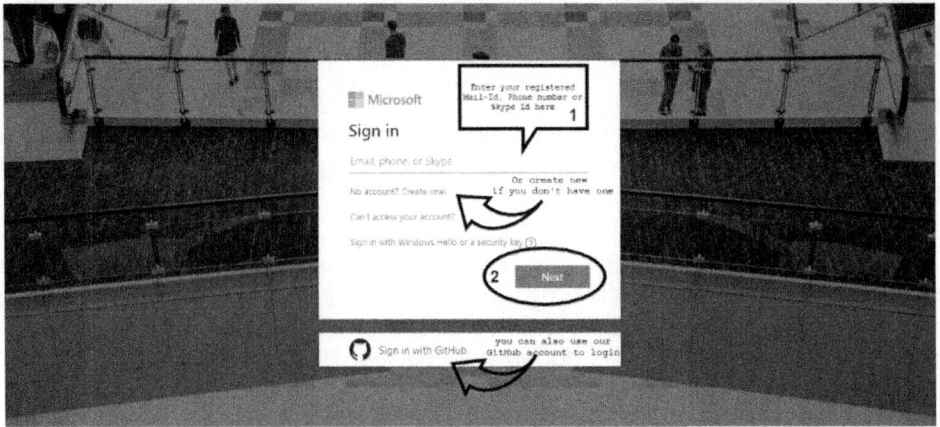

Step 3: After logging in to your Microsoft Account. You will be redirected to the next page, as shown below. Here you need to fill the required fields, and they will ask for your credit card number to verify your identity and to keep out spam and bots. **You won't be charged unless you upgrade to paid services.**

Step 4: After filling all the details, it will ask you to check the privacy and agreement. Click the checkbox and then click on Sign up.

Step 5: Your free Account is created, and you will be redirected to the Azure homepage, as shown in the figure below. You can take a tour of Azure services.

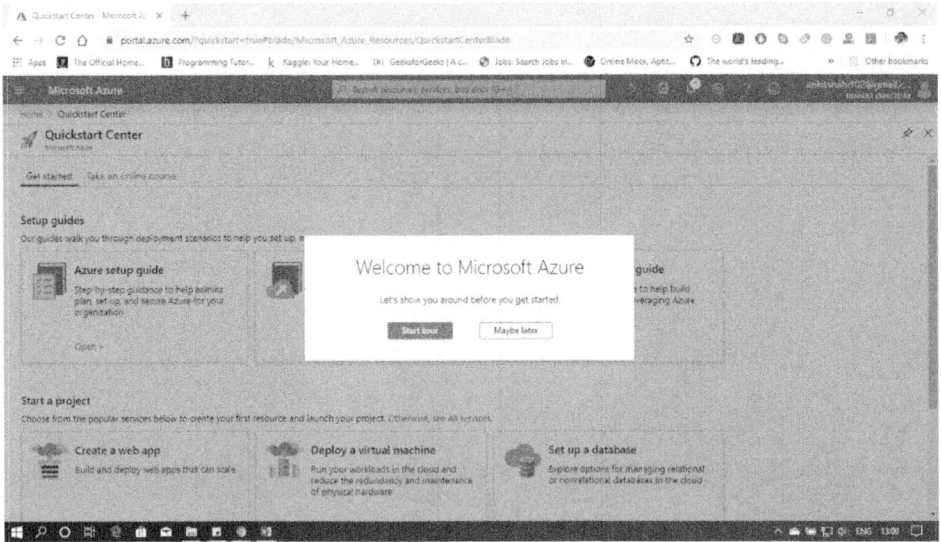

When you log-in to Azure for the first time. The Azure portal looks similar to the picture given below. We will see popular tools and services on the homepage.

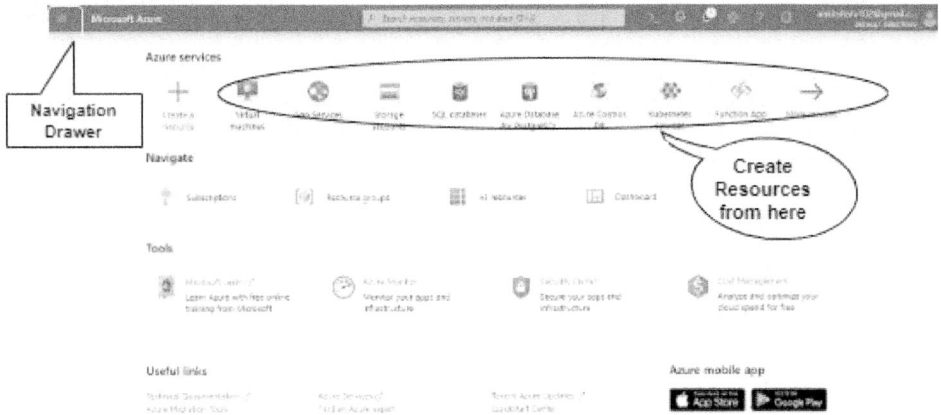

Creating a Resource

To create a resource, you can select any resource form the homepage.

Or, if you want to create another resource that is not on the homepage, you can browse the Navigation Drawer on the top left corner of the screen.

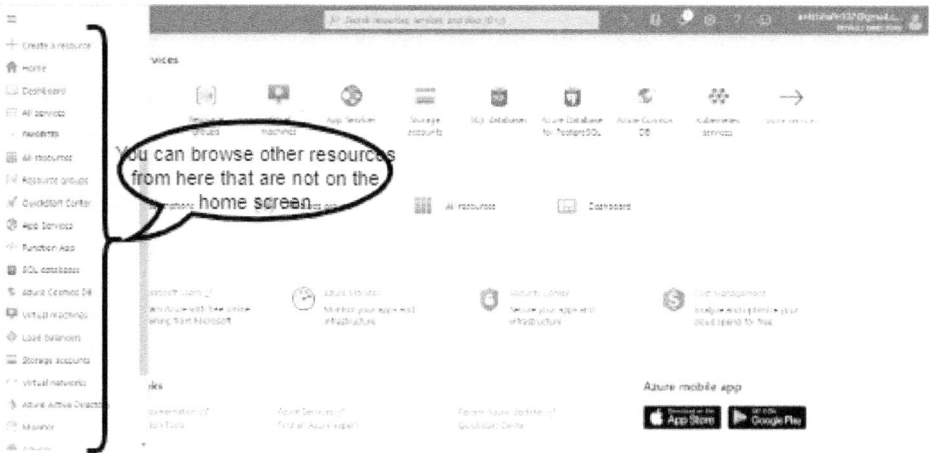

In case if you can't find the right resource in the navigation drawer, you may click on "All Service" in the navigation drawer, and the following window will appear with all the services available in Azure.

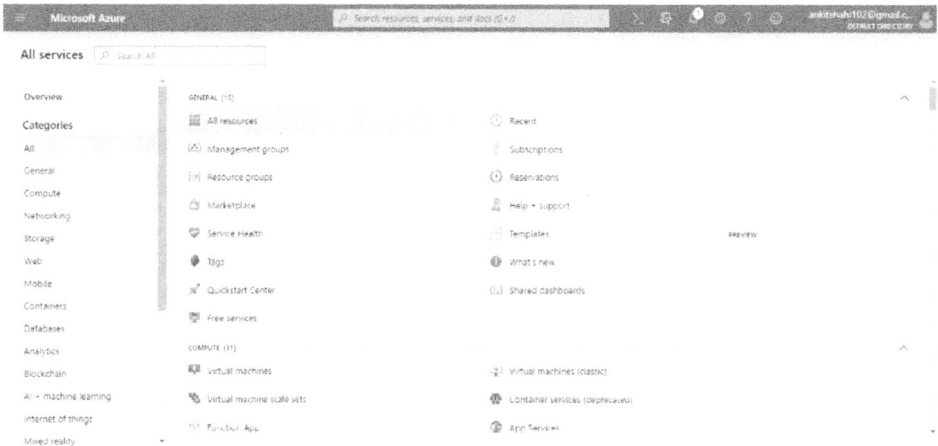

There is one more way to create a resource. Click on "Create a resource" and then type the desired resource name in the search box.

This portal not only includes the services provided by Azure but also includes service provided by the third-party providers on the platform of

Azure. They were using CPU or virtual machines of Azure and deployed their platform on it and offering that platform as a Service to you on a pay as you go basis.

Resource Group: A container that holds the related resources for an Azure solution. It can include all the resources for the solution or include only those resources which you want to manage as a group. Resource groups are containers of resources that have a common lifecycle or share an attribute such as "all SQL servers" or "Application Attendance".

Creating a Resource Group

Step 1: Hover your cursor over the "Resource groups" button inside the navigation drawer, then click on "Create" in the appeared pop up.

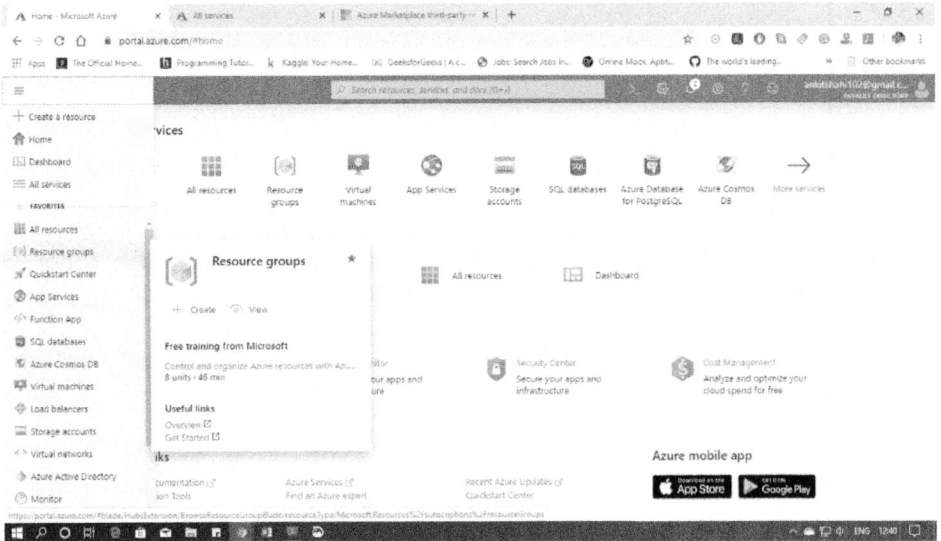

Step 2: In the next window, you have to fill the "Subscription" type, Resource group name, and Region. Then click review + create or next (to add tags).

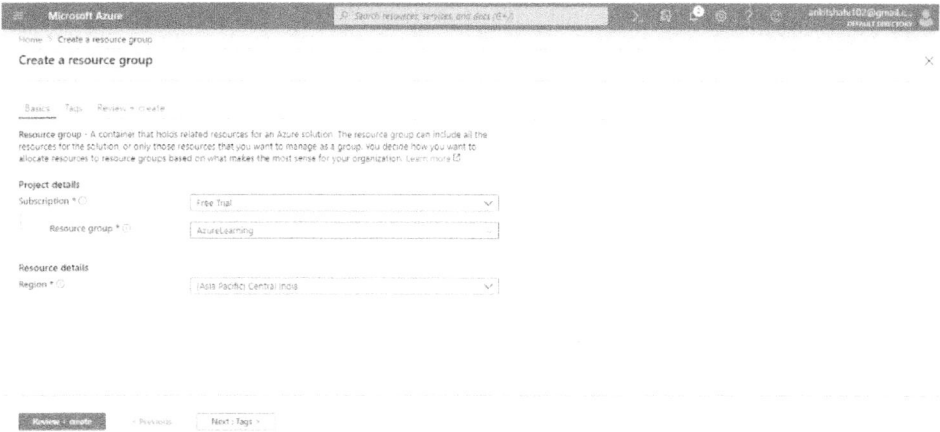

Step 3: You are now on the Tags window, where you can create a tag to organize Azure resources by categories logically. We have to give it a name and value. Click Next

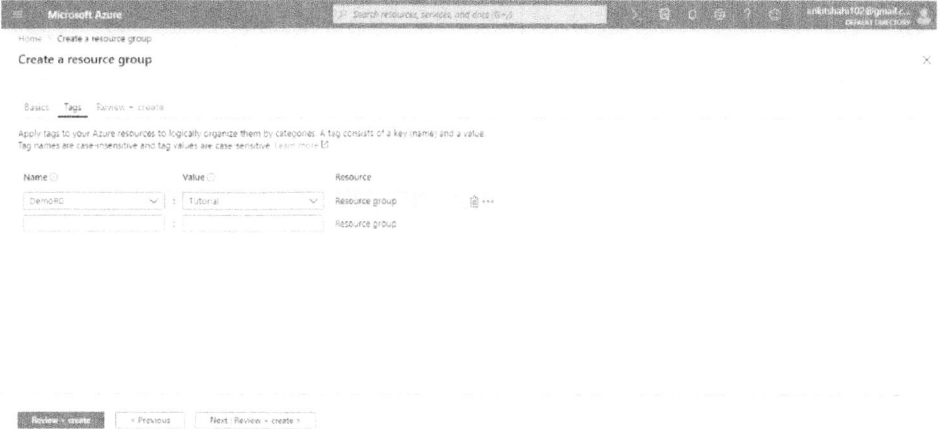

Step 4: You are now on the Review + Create window, Check the details shown below, if they are correct, then click on create.

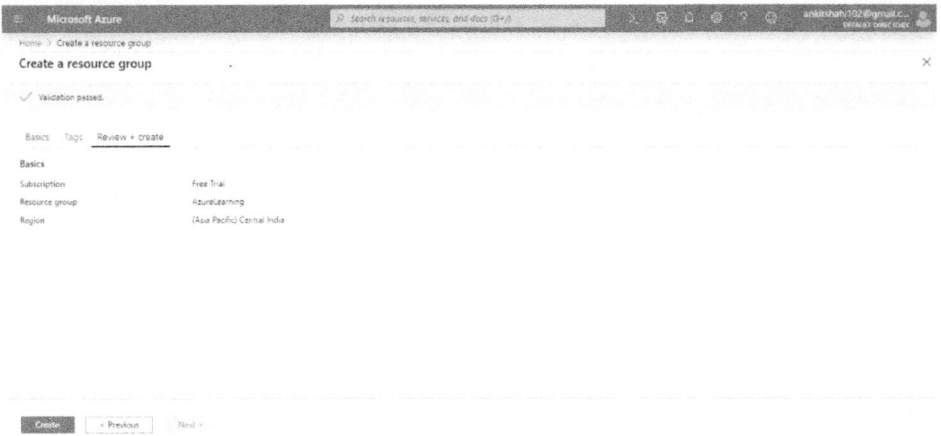

Step 5: You will be redirected to the homepage, and a notification will appear showing the Resource group is created.

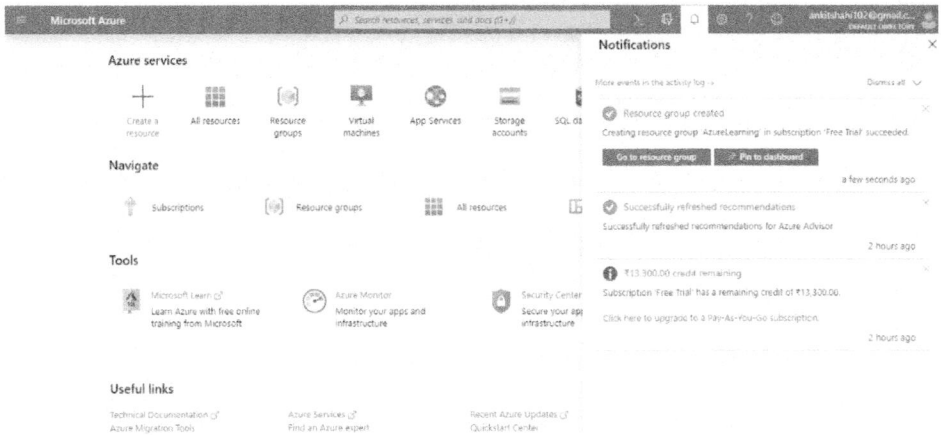

Step 6: Click on Go to Resource group to view the resource group window.

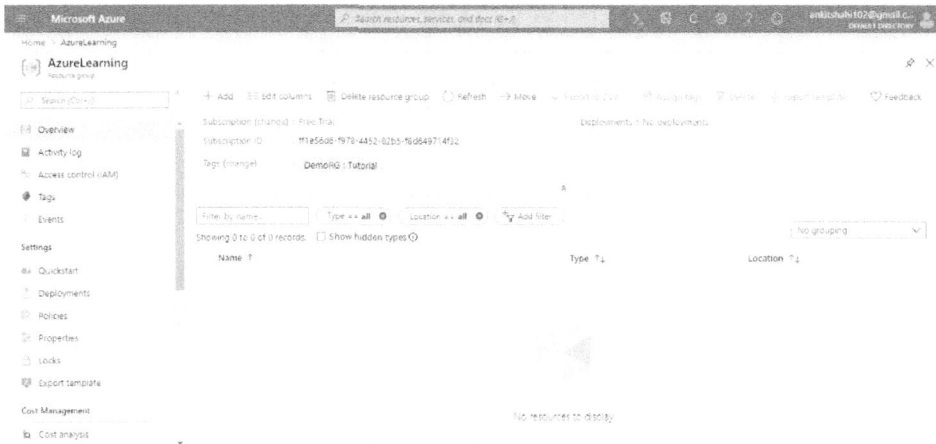

Let's have a quick look at the menu item of this page, but remember it may be different for different resources. For example, the configuration setting will be different for VMs as compared to databases.

1) **Overview:** On the overview pan, we can see all the resources that belong to that resource group and also some Metadata of the resource groups such as to which subscription it belongs to, any tags associated with it, what deployments had been carried out, etc.

2) **Activity log:** It provides administrative activity data that has been carried out on that particular resource. So, in this case, we create a resource group. Hence we have one update resource succeeded. So when we click on it, we can see the Metadata associated with it, and when we click on JSON, we can see what operations have been carried-out (See figure below).

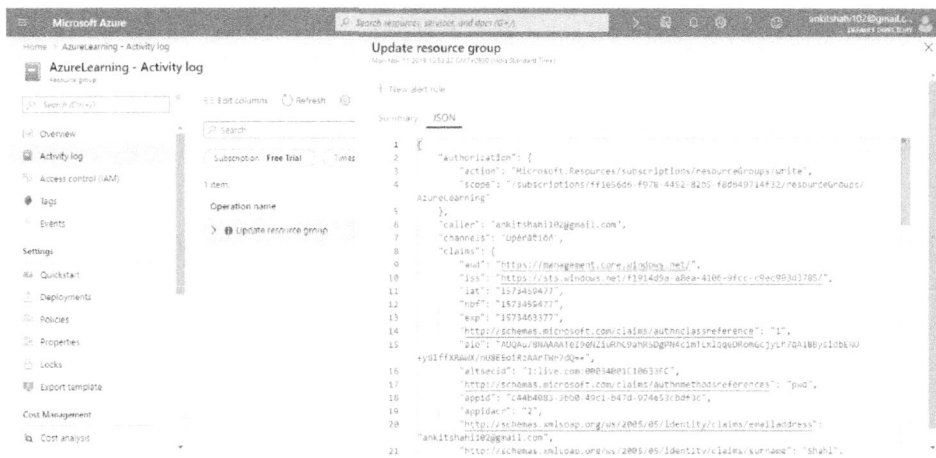

3) **Access control (IAM):** If we want to delegate access to any resource to somebody else, then we can assign a contributor role or owner role to any resource group to somebody. And the details of this role-based access control we can find on the Security Services page of this book.

4) **Tags:** We can assign Tags to any resource in Azure to classify them into categories.

5) **Events:** Any events that are happening in any particular resource group, we can subscribe to those events and do something with it. For example, a virtual machine has been started or stopped. In that case, we can capture the event and send an email to somebody.

6) **Deployment:** We can see any implementations that happened here.

7) **Policies:** We can create and view policies here.

8) **Cost Management:** We can view the resource cost here.

9) **Monitoring:** We can set alerts, see the metrics associated with this resource group, diagnostic settings, and so on.

Subscription

To view subscription. Go on the search box and type subscription and click on "Subscription," as shown in the figure below. You can see the subscribed services here.

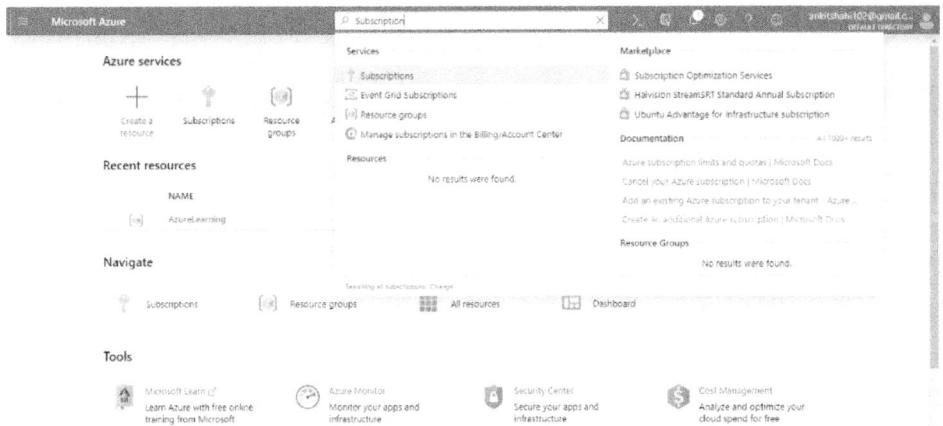

We are subscribed to the Free Trial here. See the figure below.

[20]

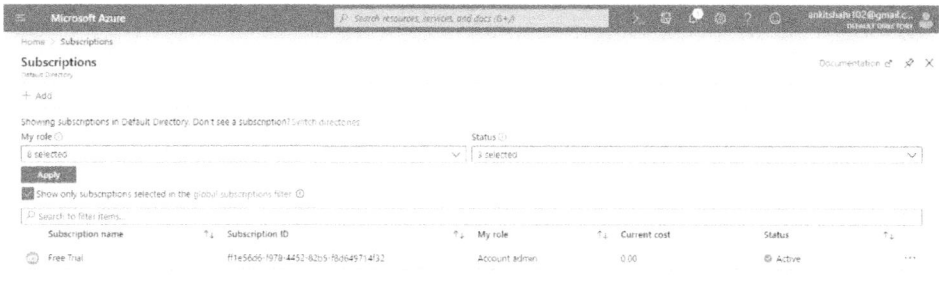

When you click on the subscription, you can view all the details, which includes the subscription name, Cost, ID, to which directory it belongs to, and billing period.

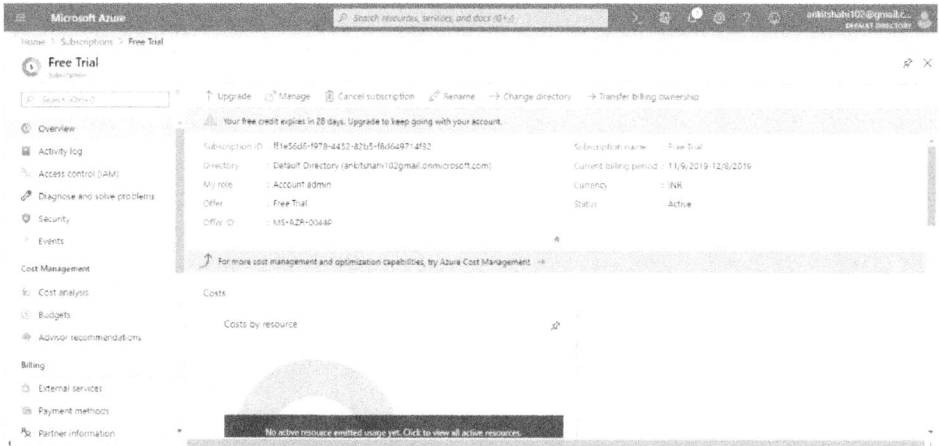

There is an integrated management portal, which we need to see.

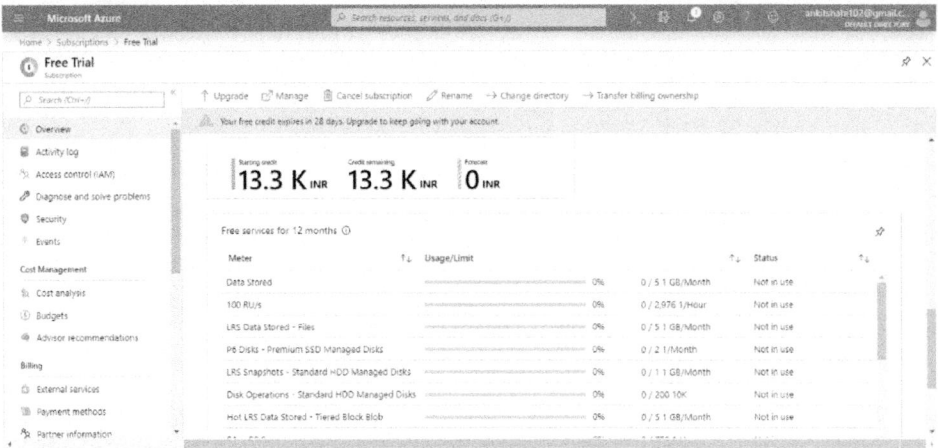

It is a portal where we can see all the subscriptions we have.

So we can see here, we have one free subscription. If we click on Free Trial, we can see all the costs we have incurred, and we can see the billing history, and also we can set alerts. For example, if the cost is crossing a certain limit, then we will get alerted.

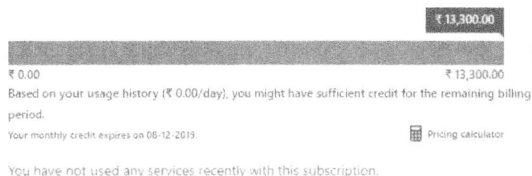

HOME   PRICING   DOCUMENTATION   DOWNLOADS   COMMUNITY   SUPPORT   ACCOUNT

Portal →

subscriptions   marketplace   profile   preview features

Summary for **Free Trial**

OVERVIEW   BILLING HISTORY

ⓘ Your Free Trial expires in 27 day(s). Click here to automatically convert to Pay-As-You-Go and avoid service disruption.

ⓘ You will be redirected to the Azure portal for some features because they are deprecated in this portal. Click here to learn more about these features in the Azure portal.

₹ 13,300.00

SUBSCRIPTION STATUS

₹ 0.00   ₹ 13,300.00

Based on your usage history (₹ 0.00/day), you might have sufficient credit for the remaining billing period.

Your monthly credit expires on 06-12-2019.   Pricing calculator

You have not used any services recently with this subscription.

27
days left

₹ 13,300
credits remaining"

Upgrade now →

And on the right-hand side tile below, we have options to manage the payment methods, download the details of the uses, contact Microsoft support and change the subscription details, edit subscription details, change address, change partner, etc.

Section II: Azure Storage Service

Azure Storage Building Blocks

The fundamental building block of Azure storage service is the **Azure storage account**. The Storage account is more like an administrative container for most of the Azure storage services. All the storage services are explained below.

Azure Blob: We can have Azure Blob storage within the storage account, which is used to store the unstructured data such as media files, documents, etc.

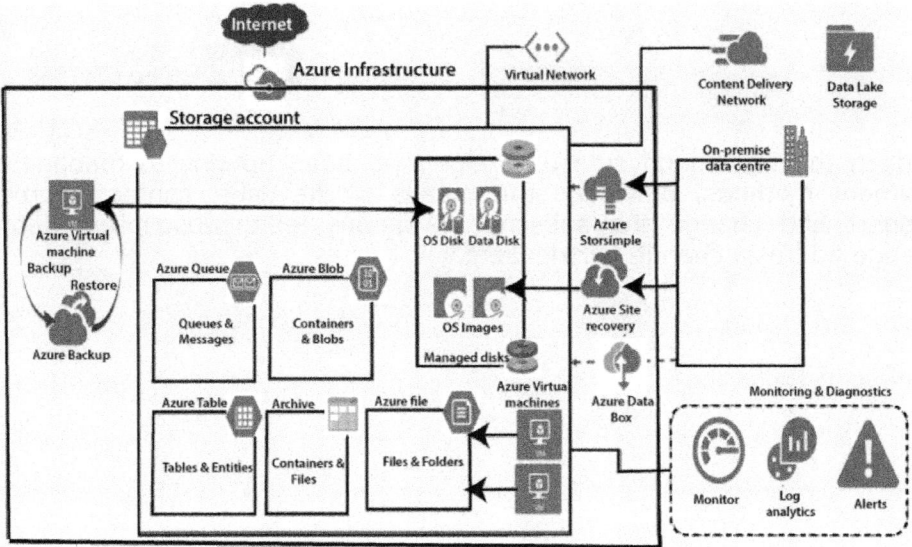

Azure file: Azure file can be used in case if we want to share files between two virtual machines, then we can create an Azure file share and access it on both of the virtual machines. We can share the data between two or more VMs.

Archive: Archive is recently introduced, and it is in preview. We can use the archive for cost optimization. So, we can move any infrequently accessed blobs or files into the archive to optimize the cost. However, once you move the data into an archive, it will take some time for the recovery of that data.

Azure Queues: It can be used to store messages.

Azure Table: It can be used to store entities. The Azure Table is a bit different from the SQL table. This is a NoSQL datastore where the schema within the table is not enforced.

And apart from all these services, there is one other key service which is:

Azure Disk Storage: Any OS disk associated with the virtual machine in Azure will get stored in a disk storage account. And also, any OS image from which this OS disk is generated will get stored as a **.vhd** file within the disk storage.

Azure Storsimple: In a hybrid cloud storage solution, Azure offers Storsimple. Storsimple is a hybrid storage solution that works at a SAN(Storage Area Network) level. It was used to be a separate company, but Microsoft brought it with them and is now offering the same services as a part of Azure and from DR (Disaster Recovery) perspective.

Azure Site Recovery: In case if we want to use Azure as a DR data-center, then we can use Azure site recovery to replicate the workloads from our on-premises data-center into Azure. Replicated workloads will be stored as images within a storage account. Whenever our on-premises data-center is down, we can run some automated scripts which will consider that recent image and build a virtual machine.

Azure Data Box: If we have terabytes of data which we want to transfer from the on-premises data center into Azure and we don't want to choose a network as an option because transferring the data over the network in terms of terabytes is not feasible. So, in that case, we can use the Azure data box. By using Azure Data Box, we can load the data into the data box and give that data box to Microsoft. Microsoft will load that data into Azure

Azure Backup: We can use Azure backup to backup the disks of our virtual machine into a recovery service vault and restore the same using that image. We have to be aware that Azure backup doesn't utilize storage to store the disk image. They are stored in the **recovery services vault**.

Azure Monitor: It can be used for the monitoring of all these services. We can use Azure Monitor for simple monitoring, and we can use **log analytics** for advance monitoring and analysis. We can also use **alerts** in case if we want to get alerted on certain things, for example, if the file share capacity is reaching its limit, then we configure it in such a way that we will get alert about the same.

CDN (Content Delivery Network): It is used for the delivery of the contents stored in the storage account. We can use a content delivery network to reduce the latency of the delivery. We'll create a CDN endpoint near to the users to reduce the latency.

Finally, the storage account will be connected to **Virtual Network**. The storage account will have a **storage firewall** where we can configure that from which virtual network you want to accept connections. So we can specify a particular IP address from where we want to allow connections or a specific subnet within a virtual network.

Azure Storage Account

An Azure Storage Account is a secure account, which provides you access to services in Azure Storage. The storage account is like an administrative container, and within that, we can have several services like blobs, files, queues, tables, disks, etc. And when we create a storage account in Azure, we will get the unique namespace for our storage resources. That unique namespace forms the part of the URL. The storage account name should be unique across all existing storage account name in Azure.

Types of Storage Accounts (All encrypted)

Storage account type	Supported services	Supported performance tiers	Supported access tiers	Replication options	Deployment model
General-purpose V2	Blob, File, Queue, Table, and Disk	Standard, Premium	Hot, Cool, Archive	LRS, ZRS, GRS, RA-GRS	Resource Manager
General-purpose V1	Blob, File, Queue, Table, and Disk	Standard, Premium	N/A	LRS, GRS, RA-GRS	Resource Manager, Classic
Blob storage	Blob (block blobs and append blobs only)	Standard	Hot, Cool, Archive	LRS, GRS, RA-GRS	Resource Manager

Note: If you want to use all storage services, we recommend you to go with general-purpose version-2, and in case if you need storage account for blobs only, then you can go with the blob storage account type.

Types of performance tiers

Standard performance: This tier is backed by magnetic drives and provides low cost/GB. They are best for applications that are best for bulk storage or infrequently accessed data.

Premium storage performance: This tier is backed by solid-state drives and offers consistency and low latency performance. They can only be used with Azure virtual machine disks, and are best for I/O intensive workload such as the database.

(So every virtual machine disk will be stored on a storage account. So, if we are associating a disk, then we will go for the premium storage. But if we are using storage account specifically to store blobs, then we will go for standard performance.)

Access Tiers

There are four types of access tiers available:

Premium Storage (preview): It provides high-performance hardware for data that is accessed frequently.

Hot storage: It is optimized for storing data that is accessed frequently.

Cool Storage: It is optimized for storing data that is infrequently accessed and stored for at least 30 days.

Archive Storage: It is optimized for storing files that are rarely accessed and stored for a minimum of 180 days with flexible latency needs (on the order of hours).

Advantage of Access Tiers:

When a user uploads the document into the storage, the document will initially be frequently accessed. During that time, we put the document in the hot Storage tier.

But after some time, once the work on the document is completed. Nobody generally accesses it. So it will become infrequently accessed document. Then we can move the document from Hot storage to Cool storage to save the cost because cool storage is built based on the number of times the document is accessed. Once the document is matured, i.e., once we stopped working on that document, the document becomes old. We rarely refer to that document. In that case, we put it in cool storage.

But for six months or one year, we don't want the document to be referred to in the future. In that case, we will move that document to archive storage.

So hot storage is costlier than cool storage in terms of storage. But cool storage is more expensive in terms of access. Archive storage is used for archiving the documents into storage, which is not accessed.

Azure Storage Replication

Azure Storage Replication is used for the durability of the data. It copies our data to stay protected from planned and unplanned events, ranging from transient hardware failure, network or power outages, and massive natural disasters to man-made vulnerabilities.

Azure creates some copies of our data and stores it at different places. Based on the replication strategy.

LRS (Local Redundant Storage): So, if we go with the local-redundant storage, the data will be stored within the data center. If the data center or the region goes down, the data will be lost.

ZRS (Zone-Redundant Storage): The data will be replicated across data centers but within the region. In that case, the data is always available within the data center, even if one node is not available. OR we can say that the data will be available also if the entire data center goes down because the data is already copied in another data center within the region. However, if the region itself is gone, then you will not get the data access.

GRS (global-redundant storage): To protect our data against region-wide failures. We can go for global-redundant storage. In this case, the data will be replicated in the paired region within the geography. And in case if we want to have read-only access to the data that is copied to another region, then, in that case, we can go for **RA-GRS (Read Access global-redundant storage)**. We can get different things in terms of durability, as we can see in this table below.

Scenario	LRS	ZRS	GRS	RA-GRS
Node unavailability within a data center	Yes	Yes	Yes	Yes
An entire data center (zonal or non-zonal) becomes unavailable	No	Yes	Yes	Yes
A region-wide outage	No	No	Yes	Yes
Read access to your data (in a remote, geo-replicated region) in the event of region-wide unavailability	No	No	No	Yes
Designed to provide __ durability of objects over a given year	at least 99.999999999% (11 9's)	at least 99.9999999999% (12 9's)	at least 99.99999999999999% (16 9's)	at least 99.99999999999999% (16 9's)
Supported storage account types	GPv2, GPv1, Blob	GPv2	GPv2, GPv1, Blob	GPv2, GPv1, Blob
Availability SLA for read requests	At least 99.9% (99% for cool access tier)	At least 99.9% (99% for cool access tier)	At least 99.9% (99% for cool access tier)	At least 99.99% (99.9% for Cool Access Tier)
Availability SLA for write requests	At least 99.9% (99% for cool access tier)	At least 99.9% (99% for cool access tier)	At least 99.9% (99% for cool access tier)	At least 99.9% (99% for cool access tier)

Storage account endpoints

Whenever we create a storage account, we will get an endpoint to access the data within the storage account. So each object that we stored in Azurestorage has an address, which includes your unique account name and the combination of an account name, and service endpoint, which forms the endpoint for your storage account.

For example, if your general-purpose account name is mystorageaccount then generally the default endpoints for different services looks like:

Azure Blob storage: http://mystorageaccount.blob.core.windows.net.

Azure Table storage: http://mystorageaccount.table.core.windows.net

Azure Queues storage: http://mystorageaccount.queue.core.windows.net

Azure files: http://mystorageaccount.file.core.windows.net

In case if we want to map our custom domain for these, we can still do that. We can use our custom domain in reference to these storage service endpoints.

Creating and configuring Azure Storage Account

Let's see how to create a storage account in Azure portal and discuss some of the important settings associated with the storage account:

Step 1: Login to your Azure portal home screen and click on "Create a resource". Then type-in storage account in the search box and then click on "Storage account".

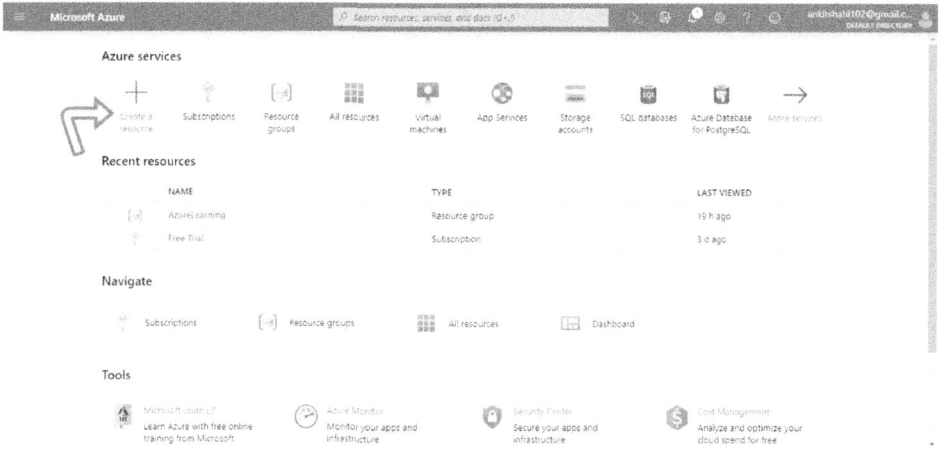

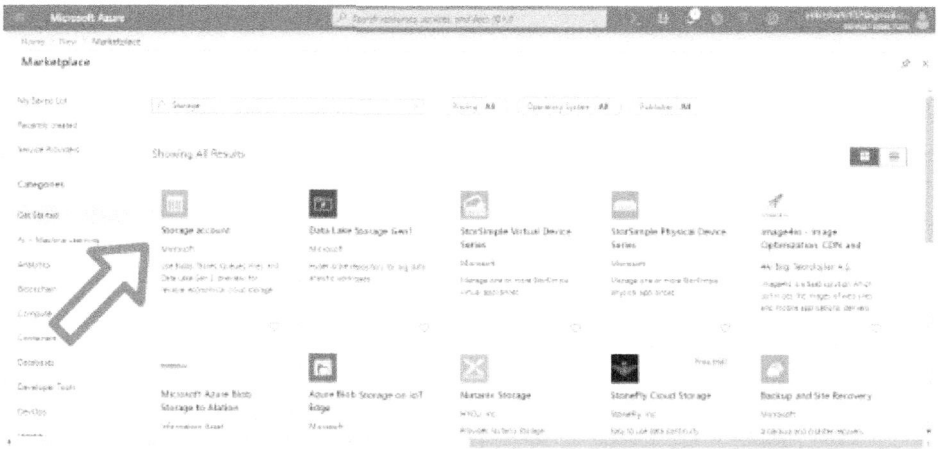

Step 2: Click on create, you will be redirected to Create a storage account window.

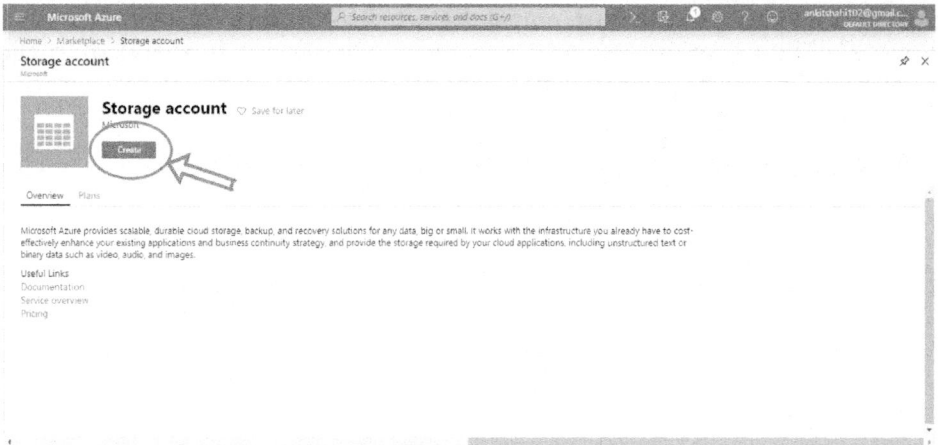

Step 3: First, you need to select the subscription whenever you are creating any resource in Azure, and secondly, you need to choose a Resource Group. In our case, the subscription is "Free Trail".

Use your existing resource group or create a new one. Here we are going to create a new resource group.

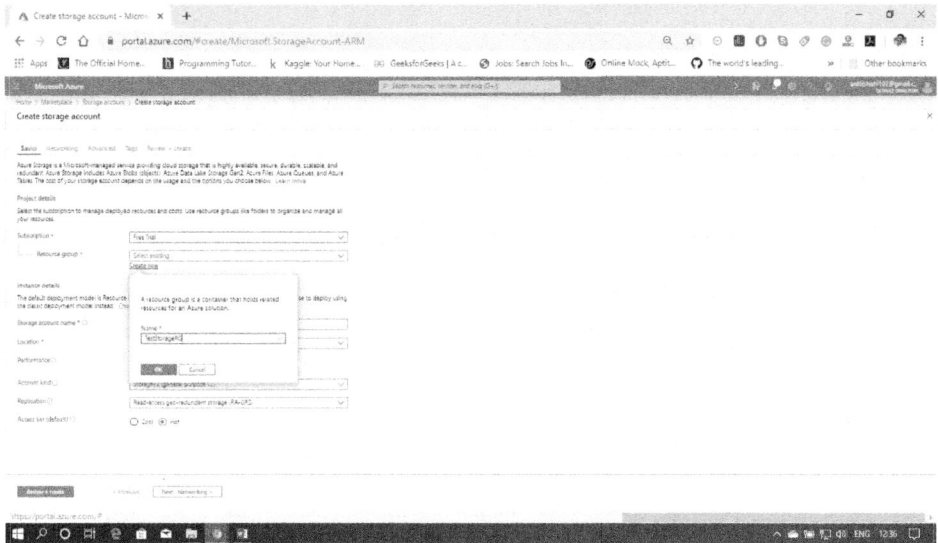

Step 4: Then, fill the storage account name, and it should be all lowercase and should be unique across all over Azure. Then select your location, performance tier, Account kind, Replication strategy, Access Tier, and then click on next.

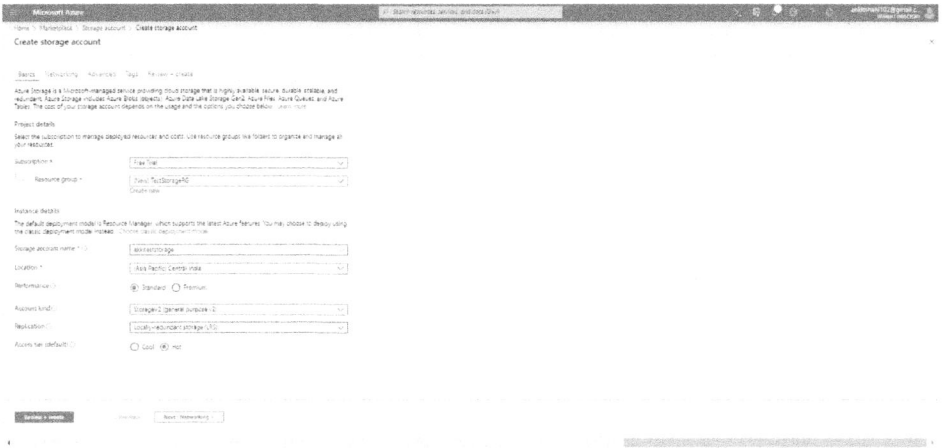

Step 5: You are now on the Networking window. Here, you need to select the connectivity method, then click next.

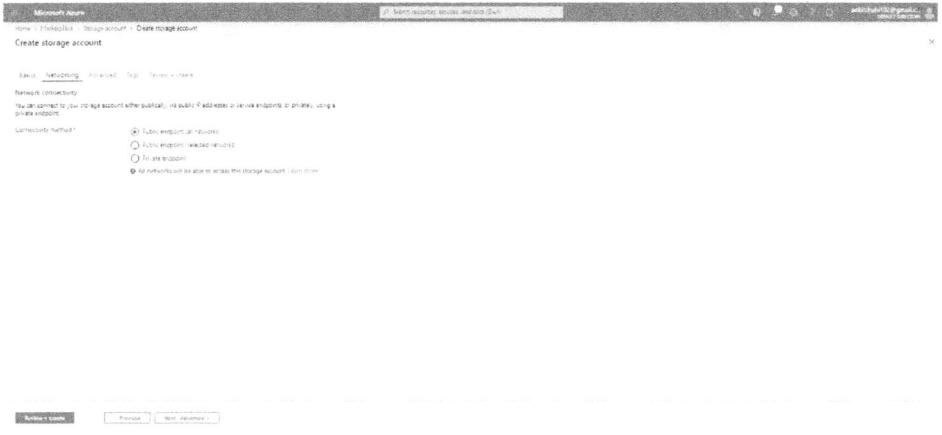

Step 6: You are now on the Advanced window were you need to enable or disable security, Azure files, Data Protection, Data lake Storage and then click next.

Step 7: Now, you are redirected to the Tags window, where you can provide tags to classify your resources into specific buckets. Put the name and value of the tag and click next.

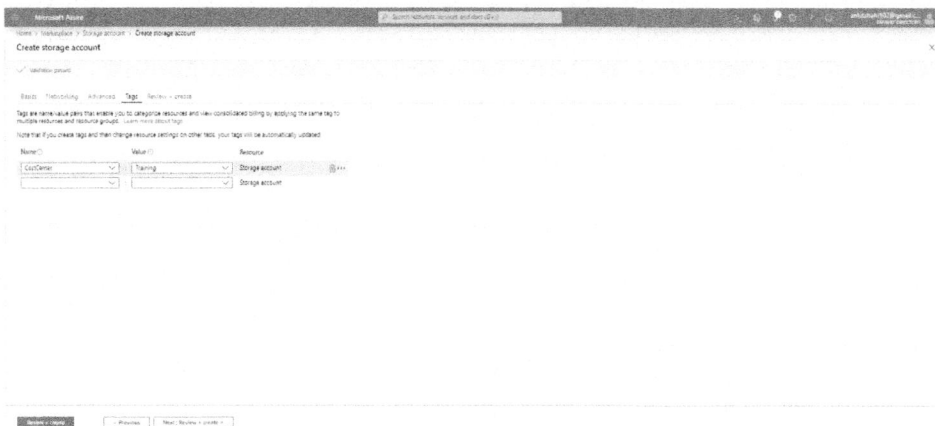

Step 8: This is the final step where the validation has been passed, and you can review all the elements that you have provided. Click on create finally.

Now our storage account has been successfully created, and a window will appear with the message "Your deployment is complete".

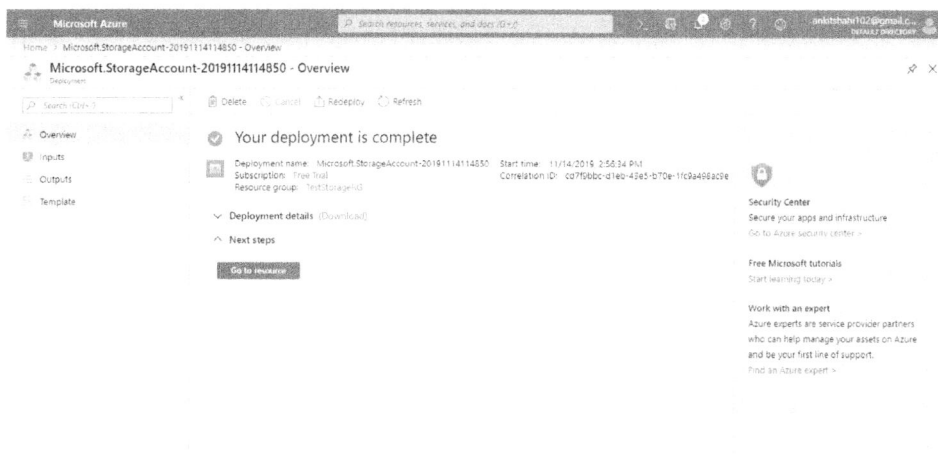

Click on "goto resource", then the following window will appear.

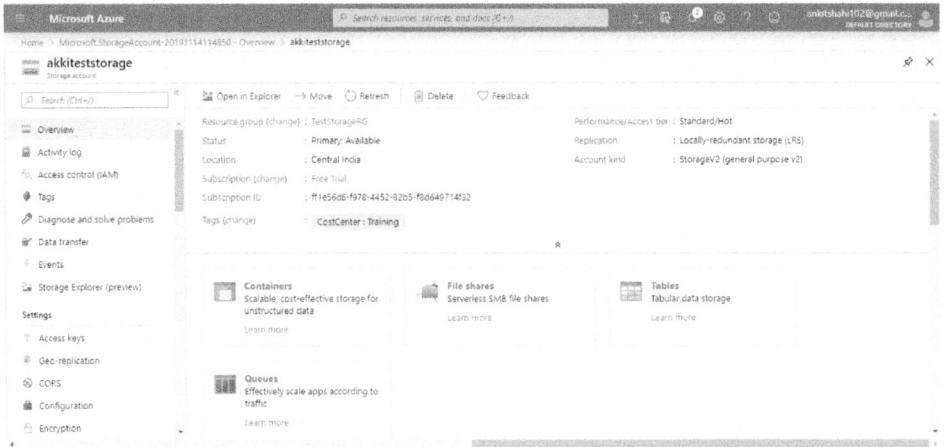

You can see all the values that you have selected for different configuration setting when creating the storage account.

Let's see some key configuration settings and key functionality of the storage account

Activity Log: We can view an activity log for every resource in Azure. It provides the record of activities that have been carried out on that particular resource. It is common for all the resources in Azure.

Access Control: Here, we can delegate access for the storage account to different users.

Tags: We can assign new tags or modify the existing tags here. We can also diagnose and solve the problems in case if we have any problems.

Events: We can subscribe to some of the events that are happening within this storage account, it can be either logic app or function. For example, a blob is created in a storage account. That event will trigger a logic app with some metadata of that blob.

Storage explorer: This is where you can explore the data that is residing in your storage account in terms of blobs, files, queues, and tables. Again there is a desktop version of this storage Explorer which you can download and connect also, but this is more of a web version of it.

Access Keys: We can use it to develop applications that will access the data within the storage account. However, we might not want to give access to this access key directly. We may wish to create SAS keys. Here, we can generate specific SAS keys for a limited period, with limited access. Then provide that SAS signature to our developers. Another way is the access keys. Access key gives blanket access. So we recommend not to give access of the access keys to anyone other than the one who created that storage account.

CORS (Cross-Origin Resource Sharing): Here, we can mention the domain name and what operations are allowed.

Configuration: If we want to change any configuration values, then there are certain things that we can't change once the storage account is created - for example, performance type. But we can change the access tier, and secure transport required or not, the replication strategy, etc.

Encryption: Here, we can specify our own key if we want to encrypt the data within the storage account. We need to click on the check box, and we can select a key vault URI where the key is located.

(SAS) Shared access signature: Here, we can generate the SAS keys with the limited access and for the limited period, and provide that information to developers who are developing applications using the storage account. SAS is used to access data that is stored in the storage account.

Firewalls and Virtual network: Here, we can configure the network in such a way that the connections from certain virtual networks or certain IP address ranges are allowed to connect to this storage account.

And we can configure advanced threat protection and can make the storage account compatible to host a static website

Properties: Here we can see the properties related to the storage account

Locks: Here, we can apply locks on the services.

So these are the different settings we can configure, and the rest of the settings are related to different services within the storage account - for example, blob, file, table, and queue.

Azure blob storage

It is Microsoft's object storage solution for the cloud. Blob storage is optimized for storing a massive amount of unstructured data, such as text or binary data.

Blob storage usages:

- ❖ It serves images or documents directly to a browser.
- ❖ It stores files for distributed access.
- ❖ We can stream video and audio using blob storage.
- ❖ Easy writing to log files.
- ❖ It stores the data for backup, restore, disaster recovery, and archiving.
- ❖ It stores the data for analysis by an on-premises or Azure-hosted service.

Azure blob storage is fundamental for the entire Microsoft Azure because many other Azure services will store the data within a storage account, inside the blob storage, and act upon that data. And every blob should be stored in a container.

Container

The container is more like a folder where different blobs are stored. At the container level, we can define security policies and assign those policies to the container, which will be cascaded to all the blobs under the same container.

A storage account can contain an unlimited number of containers, and each container can contain an unlimited number of blobs up to the maximum limit of storage account size (up to 500 TB).

To refer this blob, once it is placed into a container inside a storage account, we can use the URL, which looks like http://mystorageaccount.blob.core.windows.net/mycontainer/myblob.

Blob storage is based on a flat storage scheme. So you can't create a container within a container. Let's take an example - once we create a container like videos and if we want to differentiate between professional videos and personal videos. Then we can prefix the blob names with personnel for personal videos and professional for professional videos. The blob name will be shown as personal-video1, personal-video2 for personal videos, and for professional videos - professional-video1, professional-video2. Like this, we can create a virtual hierarchy, but we can't create a container within a container inside the Azure blob storage service.

Blob Types:

Azure offers three types of blob service:

❖ **Block blob:** It stores text binary data up-to about 4.7 TB. It is the block of data that can be managed individually. We can use block blobs mainly to improve the upload-time when we are uploading the blob data into Azure. When we upload any video files, media files, or any documents. We can generally use block blobs unless they are log files.

❖ **Append blob:** It is made up of blocks like block blobs, but are optimized for append operations. It is ideal for an application like logging data from virtual machines. For example - application log, event log where you need to append the data to the end of the file. So when we are uploading a blob into a container using the Azure portal or using code, we can specify the blob type at that time.

❖ **Page blob:** It stores random access files up-to 8 TB. Page blobs store the VHD files that backs VMs.

Most of the time, we operate with block blob and append blobs. Page blobs are created by default. When we create a virtual machine, the storage account gets created, and the disks associated with the virtual machine will be stored in the storage account. But for most of the storage solutions like we know, we are developing an application like YouTube, or we are developing a monitoring application, in that case, either we use block blobs or append blobs based on the requirement.

Naming and Referencing

The names of container and blob should adhere to some rules. Because the container name and blob name will be a part of the URL when you are trying to access them. They need to adhere to some rules which are specified below.

Container Names

❖ The name of containers must start with a letter or a number, and can contain only letters, numbers, and the dash (-) character.
❖ All the letters in a container name must be in lowercase.
❖ Container names must be 3 to 63 characters long.

Blob Names

❖ The name of blobs can contain any combination of characters.
❖ The name of blobs must be at least one character long and cannot be more than 1024 characters long.
❖ The Azure Storage emulator supports blob names up-to 256 characters long.
❖ The name of the blobs is case-sensitive.

❖ The reserved URL characters must be escaped properly.

Metadata & Snapshots

We can store some amount of information against a container or blob as metadata. It is a name-value pair associated with the container or blob. Metadata names must adhere to the name rules for C# identifiers. For example - when we are developing any video streaming application with backend as Azure blob storage, then in that case, when the user uploads a video, we want to store the user information as metadata against that video. It is very useful once we start developing an application based on blob storage.

Blob Snapshots

Snapshot is a read-only version of the blob storage. We can use snapshots to create a backup or checkpoint of a blob. A snapshot blob name includes the base blob URL plus a date-time value that indicates the time when the snapshot was created. Again if we are developing a YouTube-like application and want to retain the previous version of the video, then we can take a snapshot of it and store it once the user updates the video. So, a user like SharePoint can see the previous version of the video and the current version of the video.

To access the snapshot, we have to add a query string at the end of the URL. And a snapshot with a similar date and time when the snapshot was created.

Creating a container and adding a blob to the container.

We have already created a storage account. Now, we are going to create a container in our storage account and upload some files to it.

Step 1: Log-in to your Azure Portal and click the storage account that you have created and added to your homepage/dashboard.

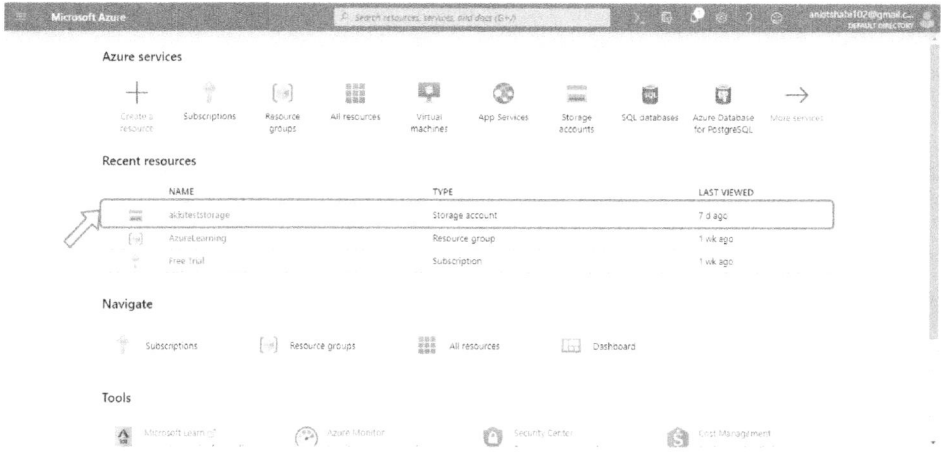

Step 2: Click on the "Containers" box, as shown in the figure below.

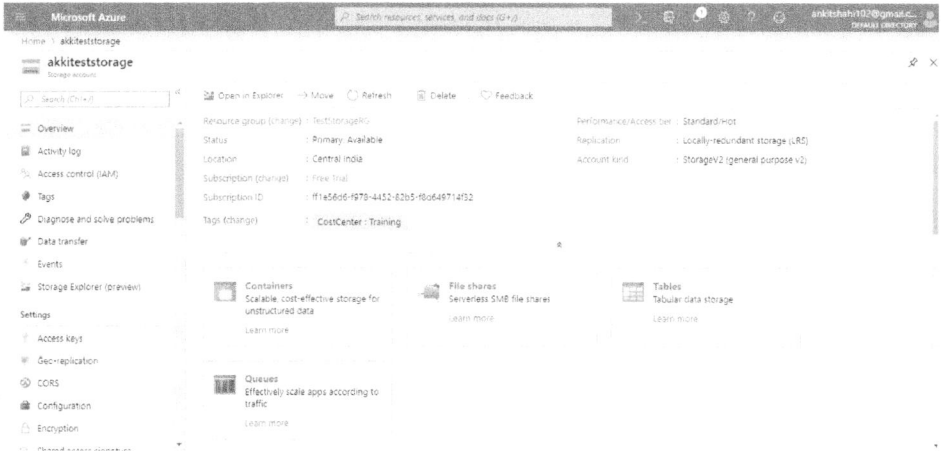

Step 3: Now, click on "+Container" tab, it will redirect you to the "container form" window.

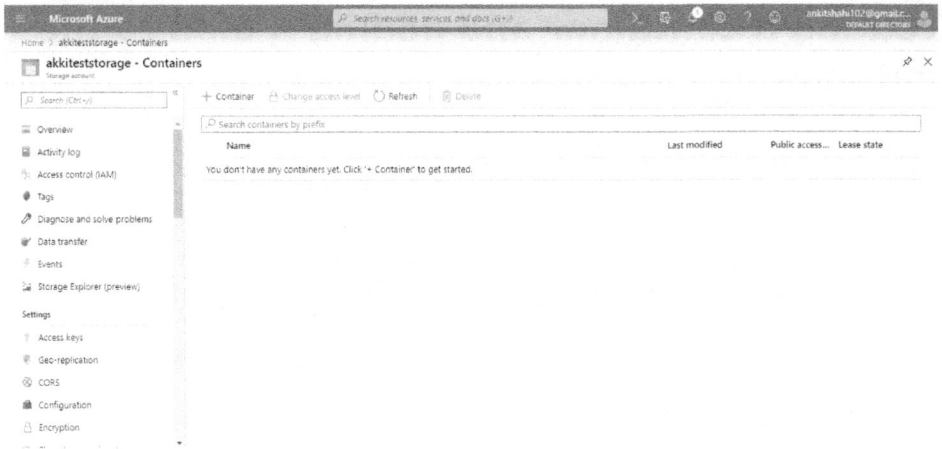

Step 4: Here, you need to assign a name to the container, and the name should be in lowercase. And in terms of access level, you can pick any from them. We are selecting blob here. Then click on, OK.

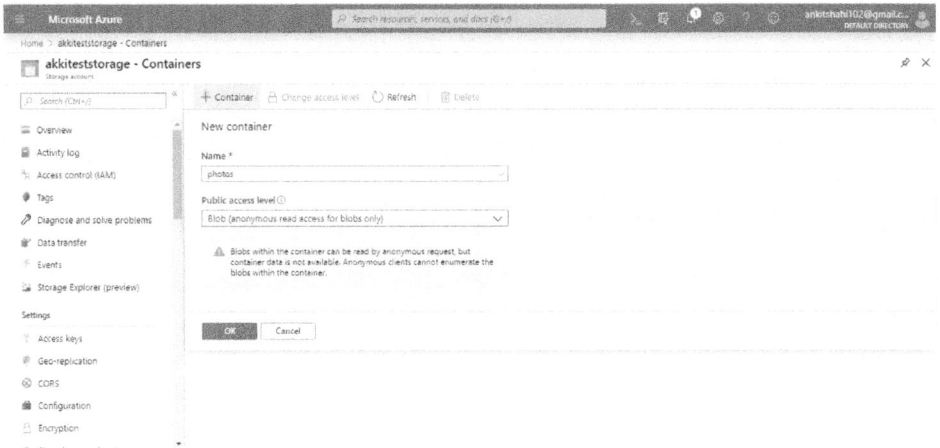

Step 5: Now, our container has been successfully created.

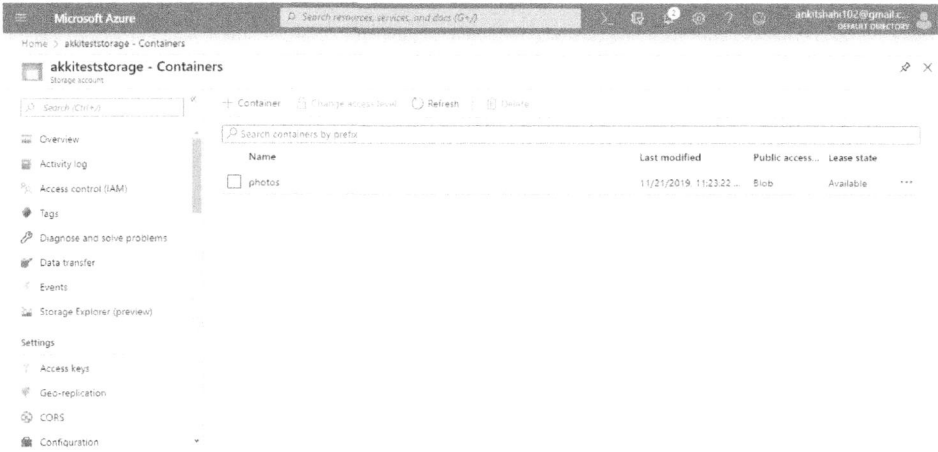

Step 6: So, if you click on the context menu, you can see the container properties and the URL using which you can access the container, last modifies Etag, and Lease status.

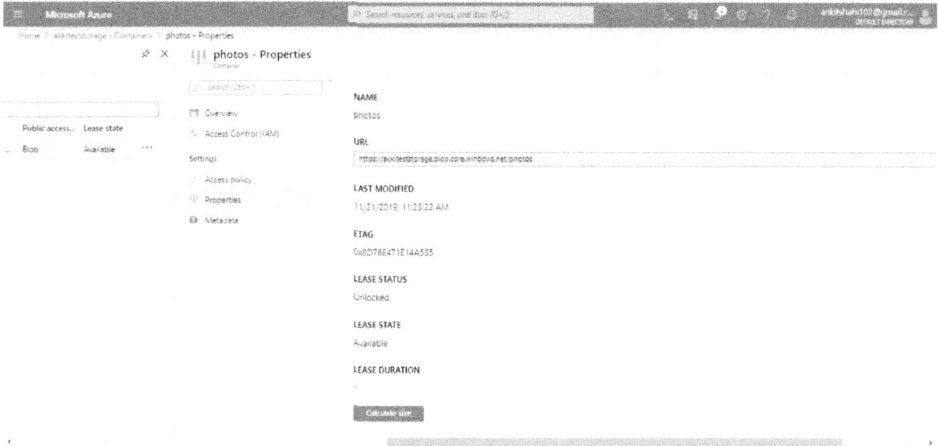

As we discussed earlier, we can have metadata at the container level and blob level also, so we can add key-value pairs in the container.

Step 7: Now, let's click on the container and upload a blob into this container.

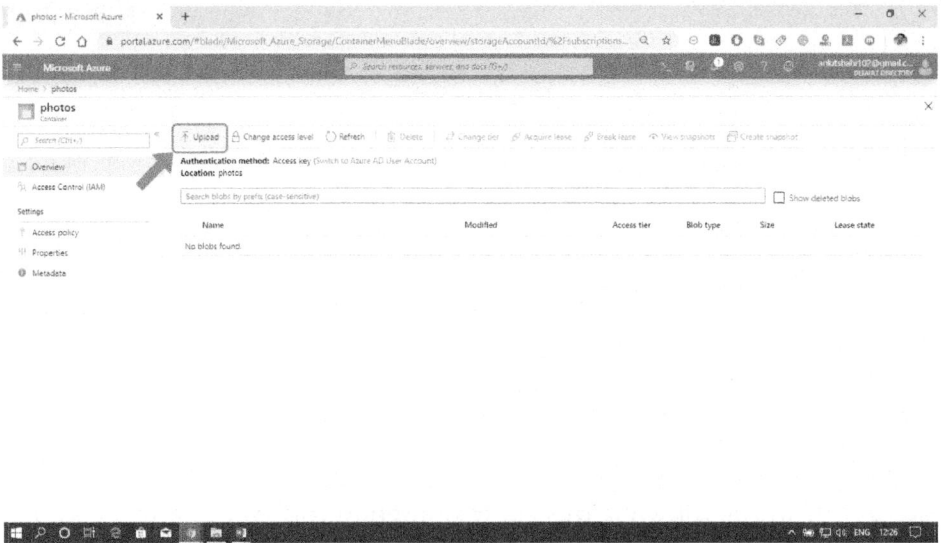

Step 8: Click on the select file option to browse the file you want to upload into the container.

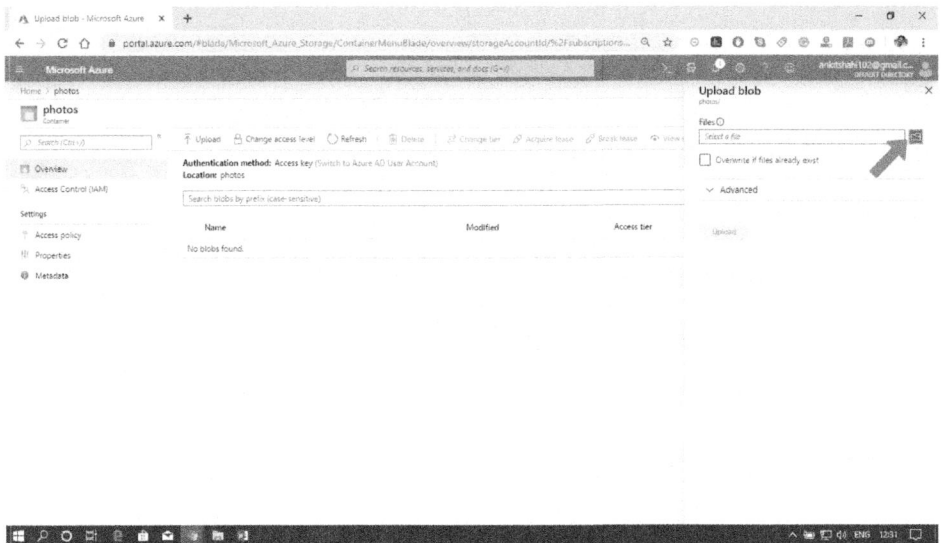

Step 9: We have selected here a JPEG file, and if we click on the advanced option, we can specify the blob type. We can define that in case if we are uploading a large file. So that the upload performance will be significantly increased because each block will be uploaded in parallel. We are thereby reducing the latency in the upload. Finally, click on the upload button.

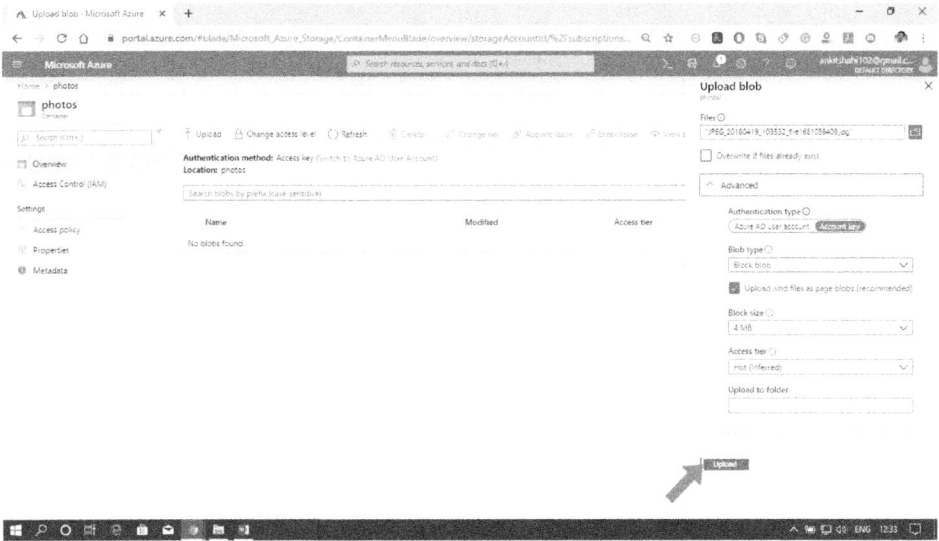

Step 10: A notification will appear once the upload will be completed. As shown in the following figure.

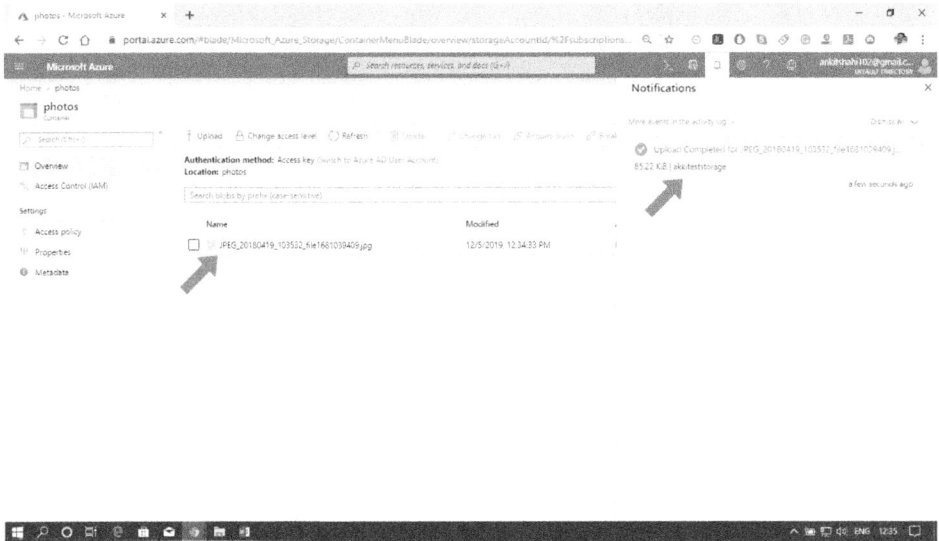

Step 11: Refresh your portal to see your file if it does not appear automatically. After that, we can see here access tier, blob type. And if we click on the menu drawer, we can see view/edit blob, download the blob, blob properties, and URL, which we can use to access this blob. We can create/view the snapshot of this particular file.

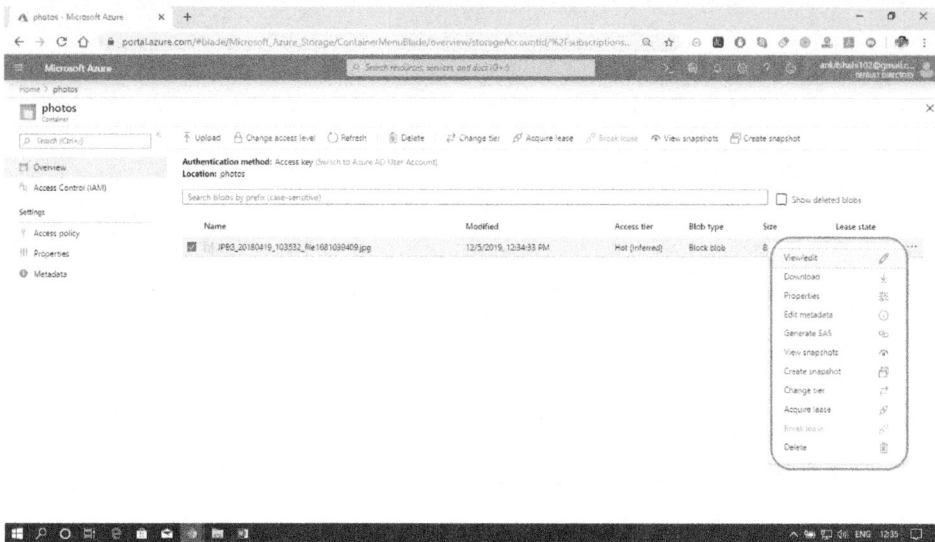

This is how we can create a container within the storage account and view the properties associated with it. And also, we can upload blobs into that container using the Azure portal.

Storage account and Blob service configuration

The first key configuration area is related to the network, which is a storage firewall and virtual networks. Every storage account in Azure has its storage firewall. Within the firewall, we can configure the following rules.

❖ A set of rules that we can configure is to allow connection from a specific virtual network. If we have an Azure virtual network, we can configure it here to enable the connections from the workloads.
❖ The second area of a rule is IP address ranges. We can specify an IP address range from where we won't allow the connections to the storage account to access the data.
❖ The third one is enabling connections from certain Azure services. So, we can specify exceptions in such a way that the connections from trusted Azure services are allowed.

We need to remember that there is a storage firewall associated with the storage account in which we can configure three types of rules.

❖ We can specify the virtual networks from which the connections are allowed.
❖ We can specify allowed IP range from where the connections are allowed.
❖ We can define some exceptions.

Custom Domain

We can configure a custom domain for accessing the block data in our storage account. The default endpoint will be the storage account name ".blob.core.windows.net". But in place of that, we can have our domain for the default storage account URL. We can configure our custom domain also. We need to specify our custom domain as "customdomain/container/myblob" to access the specific blob.

There are two fundamental limitations that we need to understand when we are using custom domains.

❖ All Azure storage docs not natively support HTTPS with the custom domains. We can currently use Azure CDN access blobs by using custom domains over HTTPS.
❖ Storage accounts currently support only one custom domain name per account. So we can use only one custom domain for all the services within that storage account.

Content delivery network

The Azure content delivery Network (CDN) caches static content at strategically placed locations to provide maximum throughput for delivering content to users. So the most crucial advantage of CDN is providing the content to the users in the most optimal way. So let's see how this works.

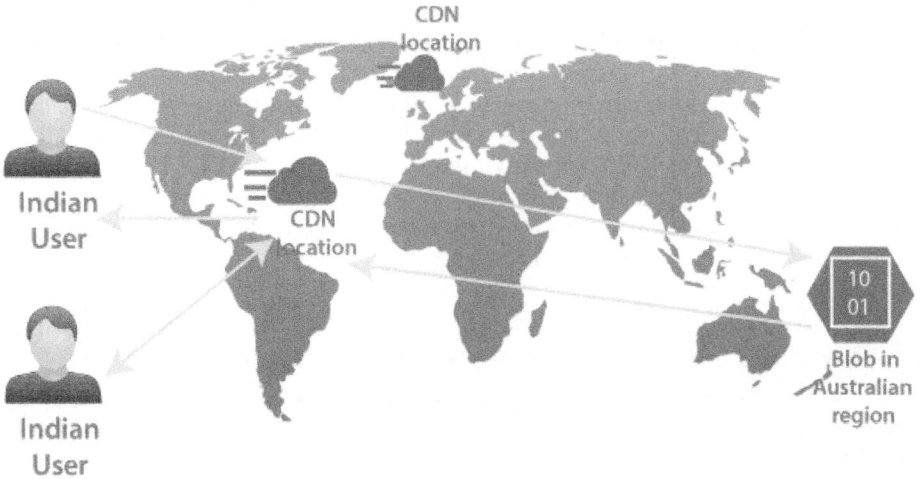

We are assuming that we have the blob storage located in the Australia region. So we have most of the users in North India and South India. In that case, we can configure a CDN profile for North India and South India. For example - let's say a North Indian user is trying to access our blob located in the Australia region. So first of all, the request goes to the CDN location. And from the CDN location, the request will further go to blob in the Australian region. For the first user, the block content will be copied to the CDN location, and then eventually delivered to North Indian users. However, when the next North Indian user tries to access that block, they will be redirected to CDN location, and the content will directly be delivered to them from that location in North India itself because the block content is already cached in CDN location.

So from the second user onwards, the content delivery latency is significantly reduced.

Other Configuration areas:

There are some different configuration areas such as performance tier, Access tier, replication strategy, secure transport required, etc.

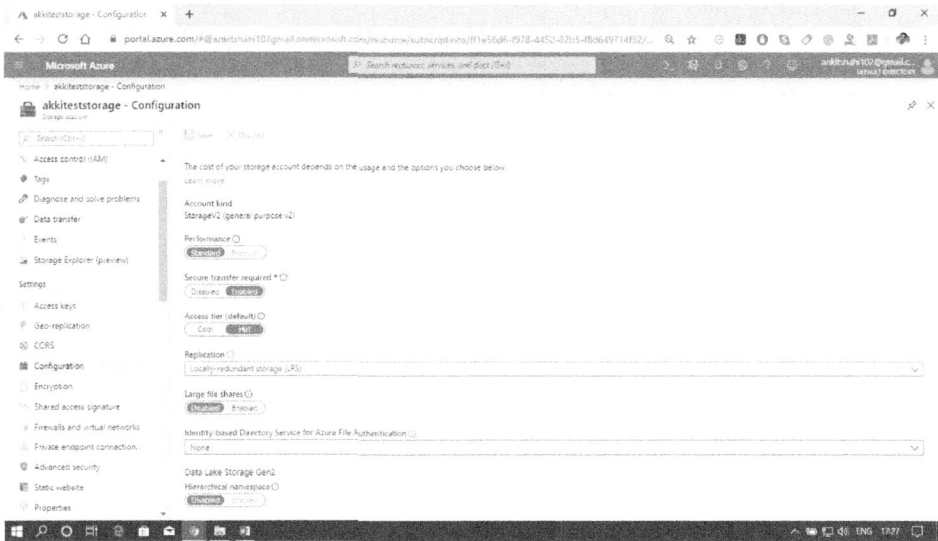

Some of them cannot be changed once the storage account is created. For example- performance tier, data lake generation 2 enabled or disabled. But we can switch on/off secure transport required, and we can change from hot to cool and cool to hot access type. We can change the replication strategy and Azure active directory authentication for Azure Files. So there are specific configuration settings that we can change.

Configuring Custom Domain for the storage account.

Let us see how to configure a custom domain for the Azure storage account, and also see some of the configuration settings we discussed above.

Step 1: Log into your storage account. Click on resource group then click on the storage account that you created

Step 2: The first thing we discussed above is firewalls and virtual networks. Here you can configure the virtual network from which you want to accept the connections to the storage account, or you can configure the IP address ranges from where you want to accept the connections and also you can specify some exceptions. For E.g., if you're going to allow the trusted Microsoft services to access this storage account to place the logs or to access the records. So, in that case, you can tick this, and also, if you want to allow read access to storage logging from any network, you can tick this. So there are some exceptions that you can make here.

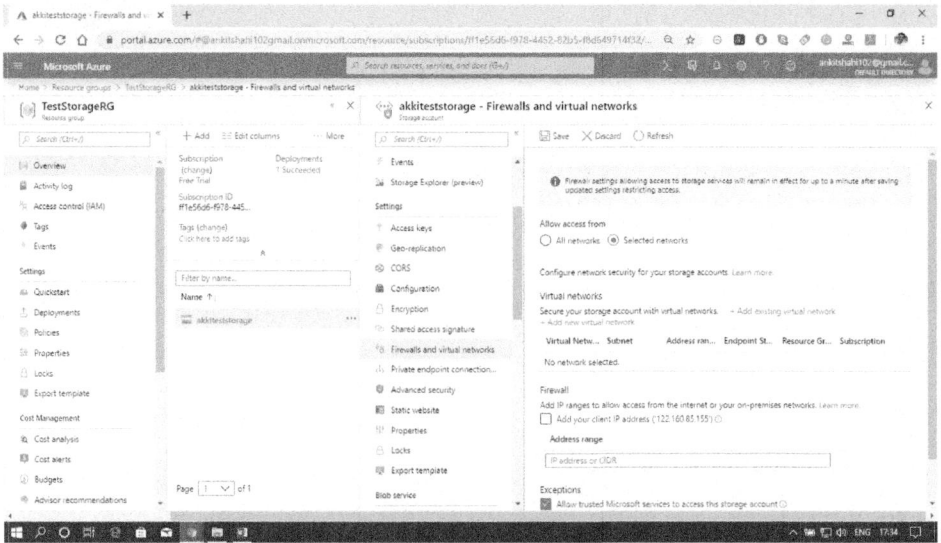

Step 3: Secondly, we discussed Azure CDN also. This is where you can configure the content delivery network endpoint. You can configure a CDN profile and map that CDN endpoint to the storage icon.

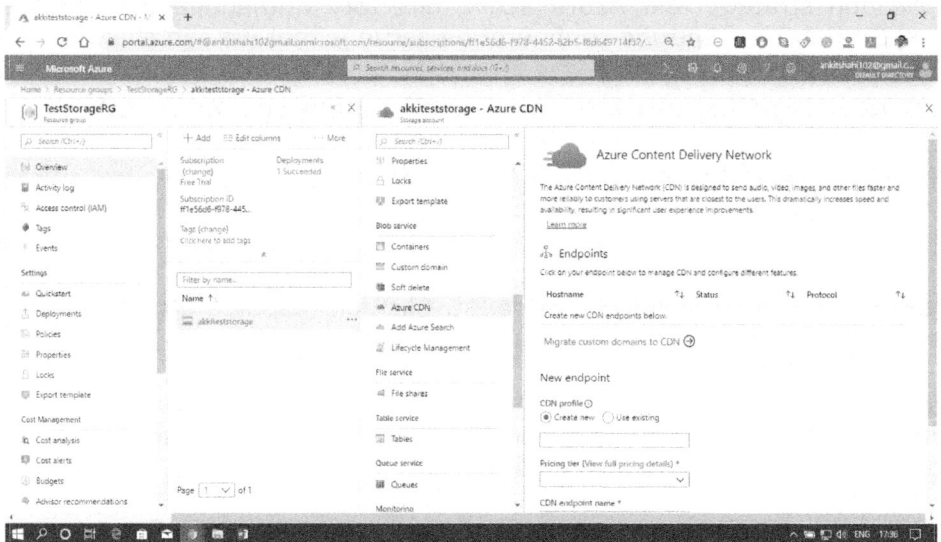

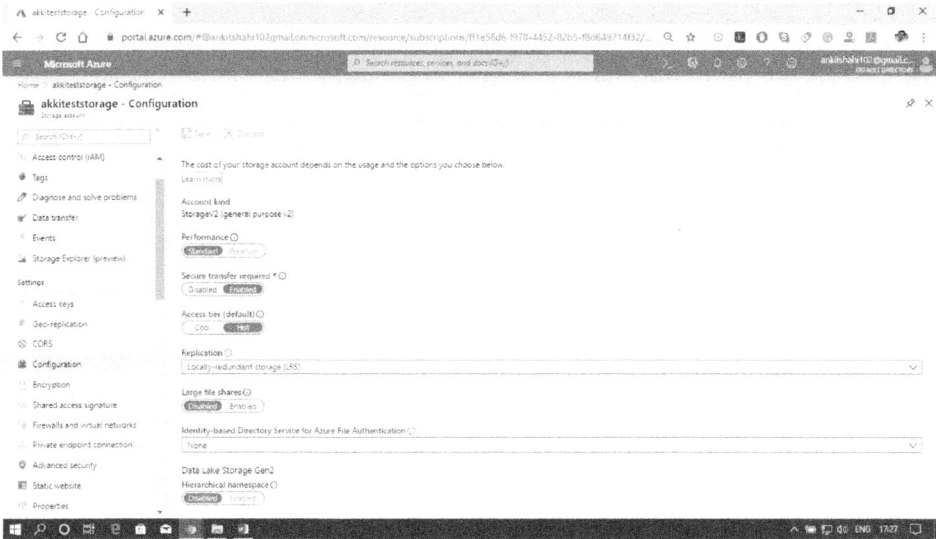

Custom domain

Step 1: Open your resource group, then your storage account, and click on the custom domain tab, as shown in the figure below.

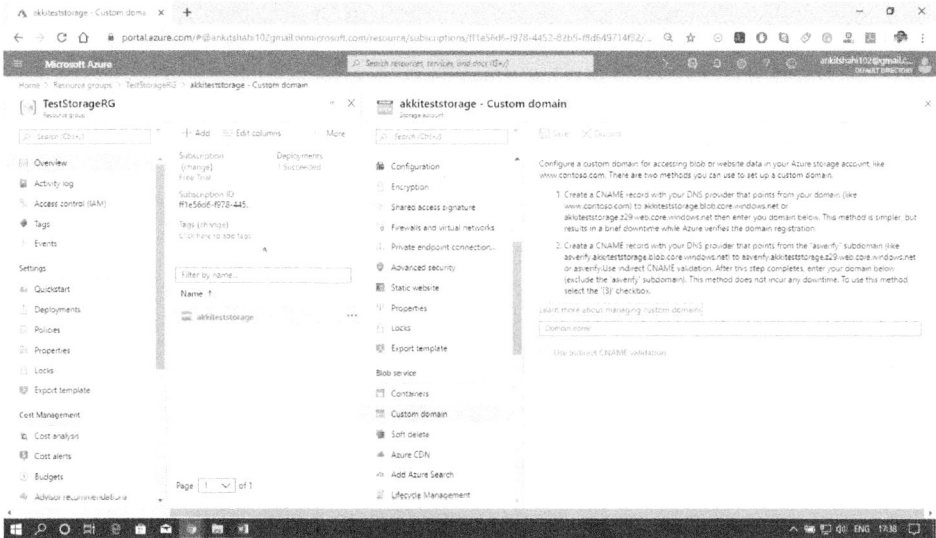

Step 2: Log in to your domain provider Web-site then click on domain DNS settings and create a Cname record.

There are two ways to do it:

You can use normal Cname, or you can use asverify also.

Step 3: Select a Cname and assign a subdomain name and after that copy and paste your blob storage link that points from your domain (like www.smaple.com) to akkiteststorage.blob.core.windows.net

Step 4: After that, fill your subdomain name inside the Domain text box in the Custom Domain window of your Storage account. Then click on Save.

Step 5: Open the browser and fill your custom domain name

You can now see the image you stored in the blob storage.

Note: Make sure your Secure transfer required is disabled as we discussed earlier, the HTTPS is not supported for Custom Domain in Azure.

Azure Storage Security

Azure storage security is divided into five major areas.

Management plane security

The management plane refers to the operation that affects the storage account itself. The way we control access to the services that affect the storage account is by using Azure active directory.

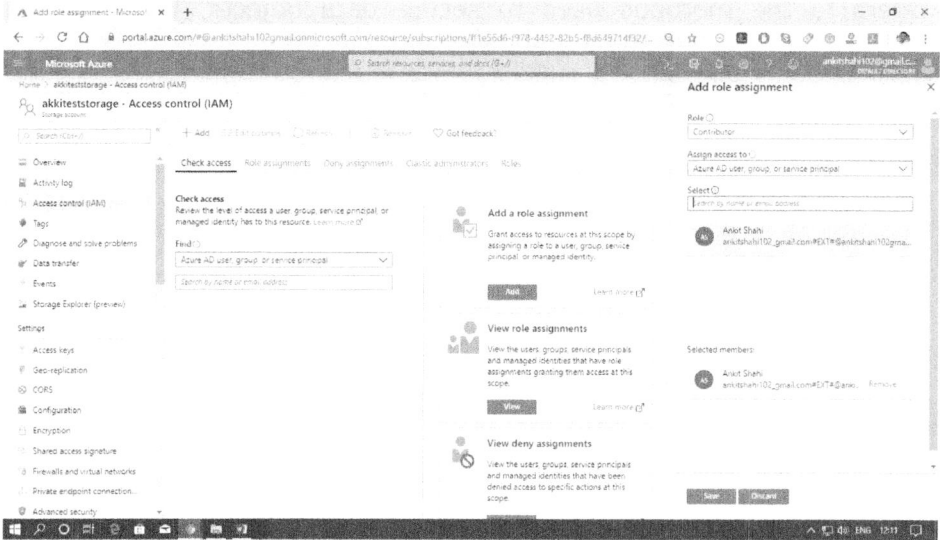

Role-based access control

❖ As we are aware that every Azure subscription has an associated Azure active directory. The Azure active directory contains users, groups, and applications. To them, we can provide access to manage resources within the Azure subscription. That resource can be a storage account, and the way we control the level of access to storage accounts is by assigning an appropriate role to the user. So we can have an owner role or contributor role or reader role that we can define.

Key Points to remember:

❖ When we are assigning a role, we can control access to the operations used to manage the storage account but not data objects in the account.

❖ However, we can give access to data objects by providing permission to read storage account keys because storage account keys enable the users to have access to data objects.

❖ Each role has a list of actions.

❖ There are some standard roles available, e.g., Owner, Reader, Contributor, etc.

❖ We can define a new custom role by selecting a set of actions from the list of available actions.

Data Plane security

It refers to the methods used to secure data objects (blobs, queues, tables, and files) within the storage account.

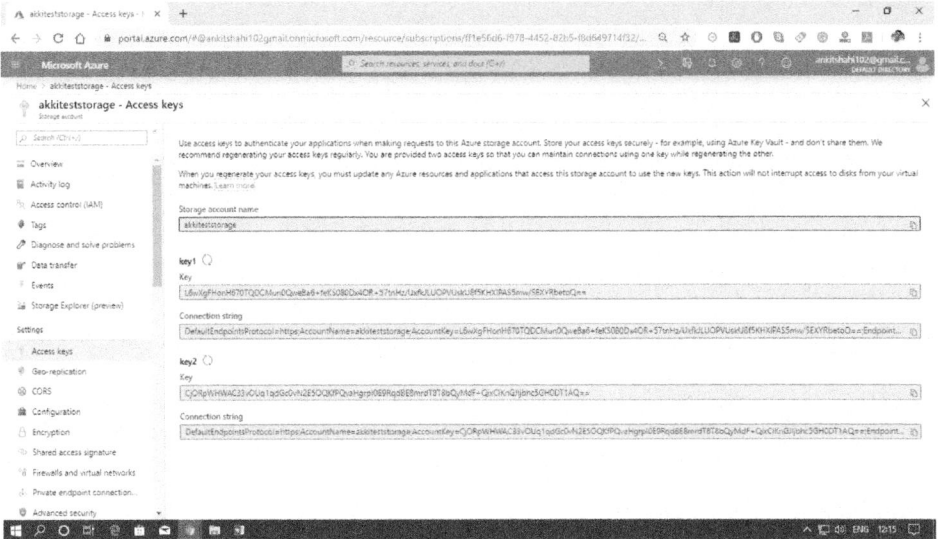

There are three ways that you can control access to the data within the storage account

❖ **Azure active directory** authorizes access to containers and queues. Azure Active Directory provides advantages over other approaches to authorization, including removing the need to store secrets in your code.

❖ **Storage account keys** provide blanket access to all data objects within the storage account.

❖ **Shared Access Signatures**, in case, if we want to provide access to certain services, for example - only to blobs, only to queues, or a combination of them. And also, if we want to control the level of access, for example - read-only, update, delete in that way, and also we wish to provide time-limited access. So we want to give access to only one year, and after that one year, we generate another SAS and present it to them for security reasons. In that case, we use shared access signatures.

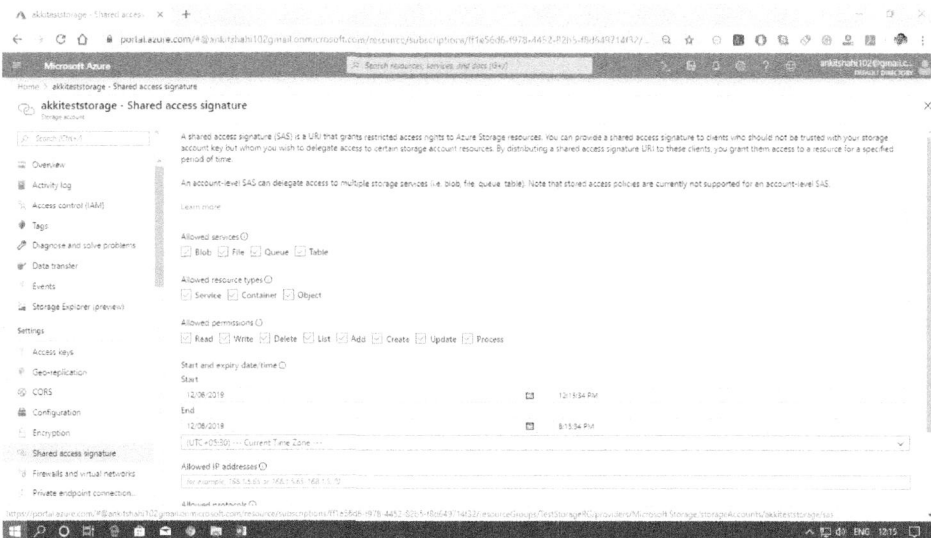

We can allow public access to our blobs by setting the access level for the container that holds the blob accordingly.

Encryption in transit

Transport level Encryption using HTTPS

❖ Always use HTTPS when using REST APIs or accessing the object in storage.
❖ If we are using SAS, we can specify that only HTTPS should be applied.

Using encryption in transit for Azure file shares

❖ 1 does not support encryption, so connections are only allowed within the same region.
❖ 0 supports encryption, and cross-region access is allowed.

Client-side encryption

❖ Encrypt the data before being transferred to Azure storage
❖ When retrieving the data form Azure, data is decrypted after it is received on the client-side.

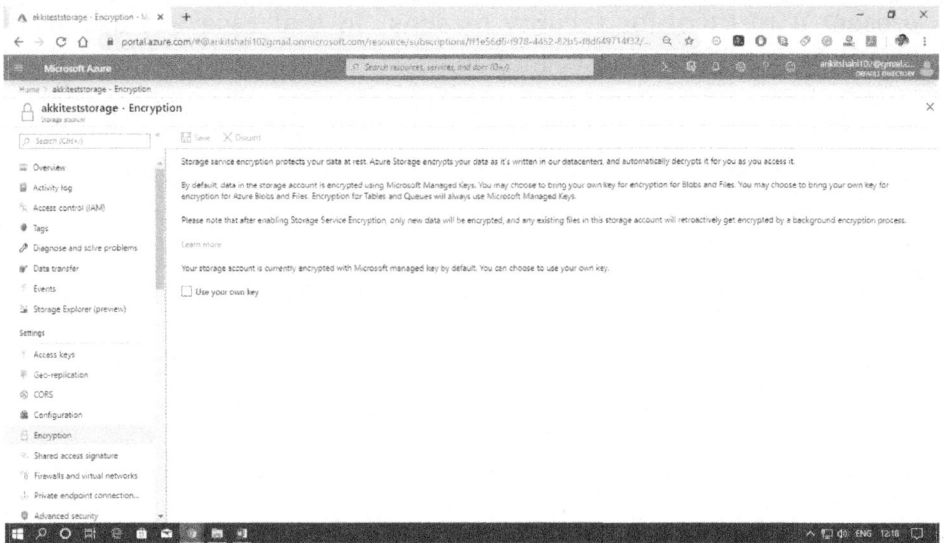

Encryption at rest

Client-side encryption

- ❖ Encrypt the data before being transferred to Azure storage.
- ❖ When retrieving the data form Azure, data is decrypted after it is received on the client-side.

Storage Service Encryption (SSE)

This is what we generally use to encrypt the data at REST is Azure storage

- ❖ It is enabled for all storage accounts and cannot be disabled.
- ❖ It automatically encrypts data in all performance tiers (Standard and premium), all deployment models (Azure Resource Manager and Classic), and all of the Azure Storage services (Blob, Queue, Table, and File). So it is blanket encryption across all Azure storage.
- ❖ We can use either Microsoft-managed keys or your custom keys to encrypt the data.

Azure Disk Encryption

This is a recommended approach from Microsoft to encrypt the disks particularly with Azure disk

- ❖ Encrypt the OS & data disks used by IaaS Virtual Machine
- ❖ You can enable encryption on existing IaaS VMs
- ❖ You can use customer-provided encryption keys

CORS (Cross-Origin Resource Sharing)

❖ When a web browser makes an HTTP request for a resource from a different domain, this is called a cross-origin HTTP request.

❖ Azure Storage allows us to enable CORS. For each storage account, we can specify domains that can access the resources in that storage account. For example, enable CORS on the mystorage.blob.core.windows.net storage account and configure it to allow access to mywebsite.com.

❖ CORS allows access but does not provide authentication, which means we still need to use SAS keys to access non-public storage resources.

❖ CORS is disabled on all services by default. We can enable it using the Azure portal or Power Shell, and we can specify the domains from where the request will come to access the data in your storage account.

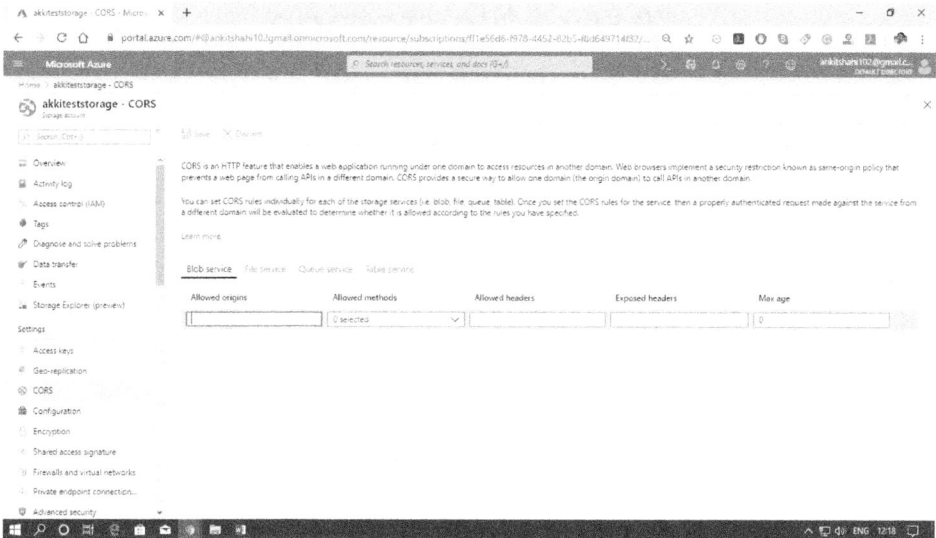

Azure File Storage Service

Azure file storage mainly can be used if we want to have a shared drive between two servers or across users. In that case, we will go for Azure file storage. In the Azure file storage structure, the first thing we need to have is an Azure storage account. Azure file storage is offered under the umbrella of the Azure storage account. And once we have created an Azure storage account, we'll create a file share.

We can create an unlimited number of file shares within a storage account. Once we create a file share, then we can create directories, just like folders, and then we can upload files into it. Once we create a file share, we can mount that on any virtual machine, whether it is in Azure or outside.

Azure file storage

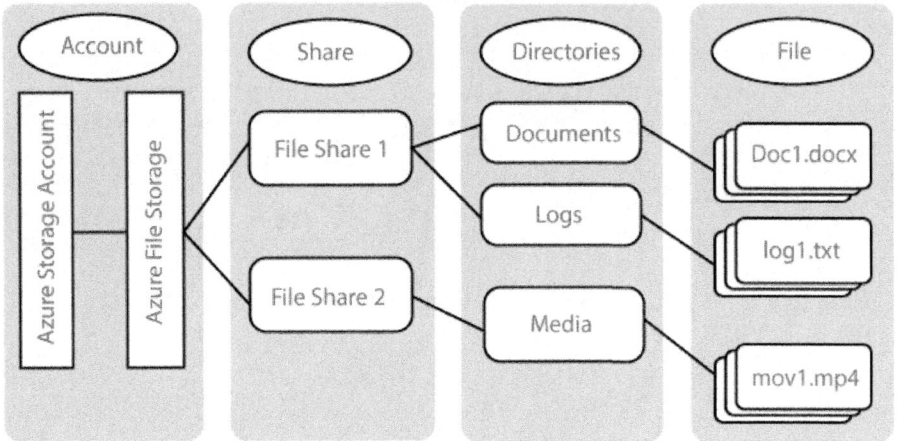

Some of the concepts related to Azure file storage:

❖ **Storage Account:** All access to Azure Storage service is done through a storage account. We need to keep in mind scalability and performance targets when we might have Azure fie storage, blob storage, tables, and queues. All of them might be sharing the same performance targets under the storage account, so the same limitations of storage account will be shared-by across all services.
❖ **Share:** A file storage Share is an SMB file share in Azure. The directories and files must be created in a parent's Share, so we can't create a directory directly in a storage account. We need file storage share, first created, and then we can create directories to upload. An

account may contain an unlimited number of shares, and a Share can store an unlimited number of files, up to 5 TB total capacity of the file share. But, in case we need more than the full capacity, then we can create another file share.

❖ **Directory:** It is an optional hierarchy of directories.
❖ **File:** A file in the share. A file may be up to 1 TB in size.
❖ **URL format:** For a request to an Azure file share made with the file REST protocol

File Storage Data Access methods

Azure files offer two, built-in, convenient data access methods that you can use separately, or in combination with each other, to access your data:

❖ **Direct Cloud Access:** Windows, MacOS, or Linux can mount any Azure file share with the industry-standard Server Message Block (SMB) protocol or via the File REST API. But if we use the SMB protocol, then we need to take care of two things. First, if we are mounting the file share on a VM in Azure, then the SMB client in the OS must support at least SMB 2.1. Secondly, if we want to mount file share on an on-premises system such as a user's workstation, then the SMB client supported by workstation must be at least SMB 3.0 with encryption enabled. And if we are mounting Azure file share within our on-premises data center, then it should be 3.0 minimum version of SMB client, but if we are installing VM in Azure, it can be 2.1.
❖ **Azure File Sync:** With Azure File Sync, Shares can be replicated to on-premises or Azure Windows Servers. Our users can access the file share through the Windows Server, such as SMB or NFS share. Synchronization of any frequently accessed files will be kept in the server endpoint, and any infrequently accessed data will be moved to Azure file share. In that way, we can get the speed of the delivery of the data to your users and, at the same time, save the storage.

Data transfer method

When we create an Azure file share, and we have a large file share already inside our on-premises data center. Then with the help of the option below, we can transfer those files.

❖ **Azure file sync:** As a part of the first sync between an Azure file share (a "cloud Endpoint") and a windows directory namespace (a "Server Endpoint"), Azure File Sync will replicate all data from the existing file share to Azure Files.
❖ **Azure Import/export:** If we have terabytes of data, which we need to transfer into Azure files. And if we start moving data using Azure file sync, it might consume all the bandwidth, or it might be a slow process that may take months. In that case, you can use Azure import/export. Microsoft will provide a hard disk to you to move all the

data in the hard drive and ship that hard drive back to Microsoft, and Microsoft will load the data from the hard-disk and into an Azure data center.

❖ **Robocopy:** Robocopy is a well-known copy tool that ships with windows and windows Server. We can use it to transfer the data into Azure files by mounting the locally shared files and then using the mounting location as a destination in the robocopy command.

❖ **AzCopy:** It is a command-line utility tool, which we can use for copying the data to and from Azure files. It can be used for blob storage also, and you can use Azcopy with simple commands. It provides excellent performance and is available for Windows and Linux.

Creating a File Storage in Azure

Step 1: Let's go into Azure storage Account, then scroll down and click on Files.

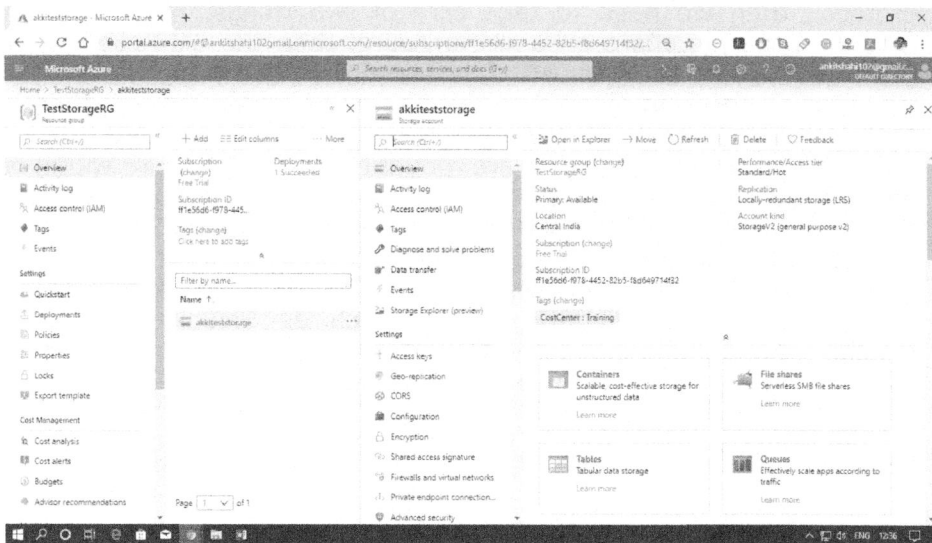

Step 2: Click on +File Share.

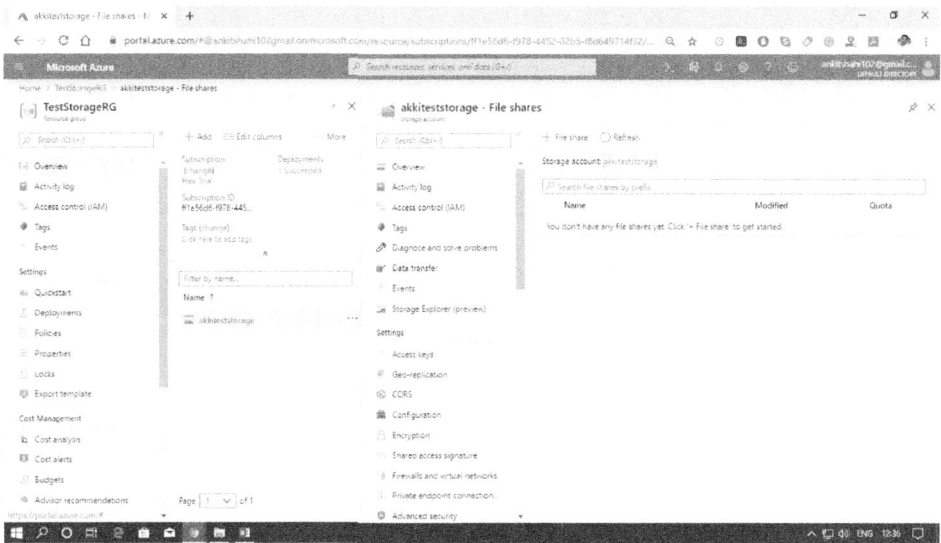

Step 3: Provide the name and Quota of the file share, then you will get a notification of the successful creation of File share.

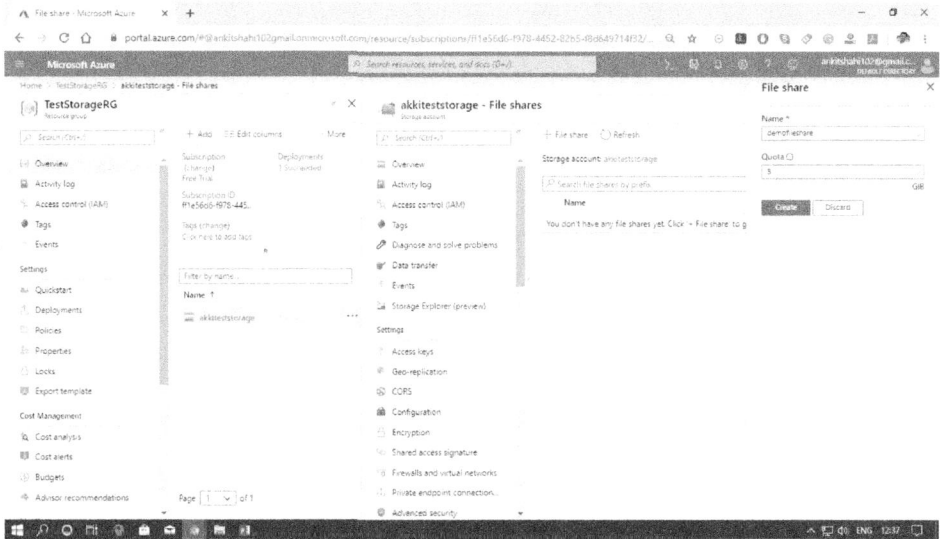

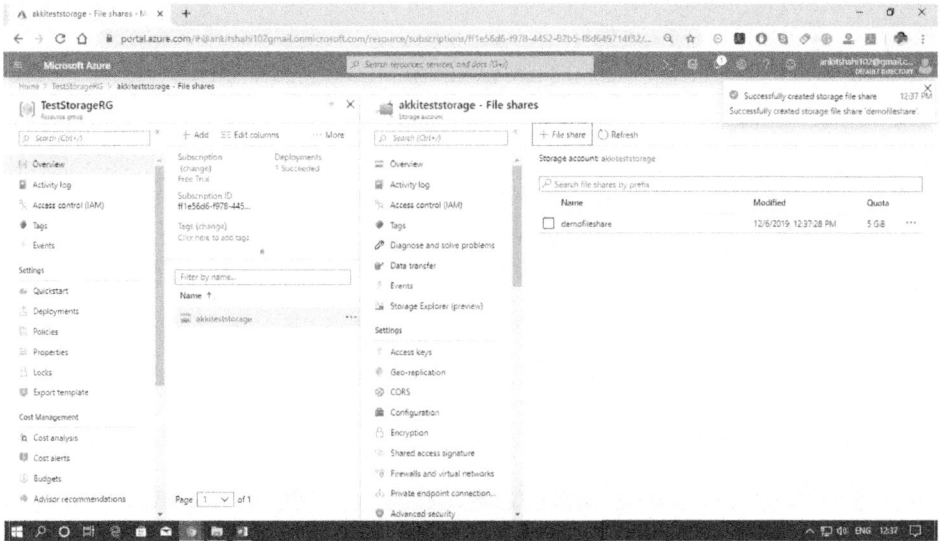

Step 4: Click on the File share property where you can see the URL and also the quota you assigned and how much it is used.

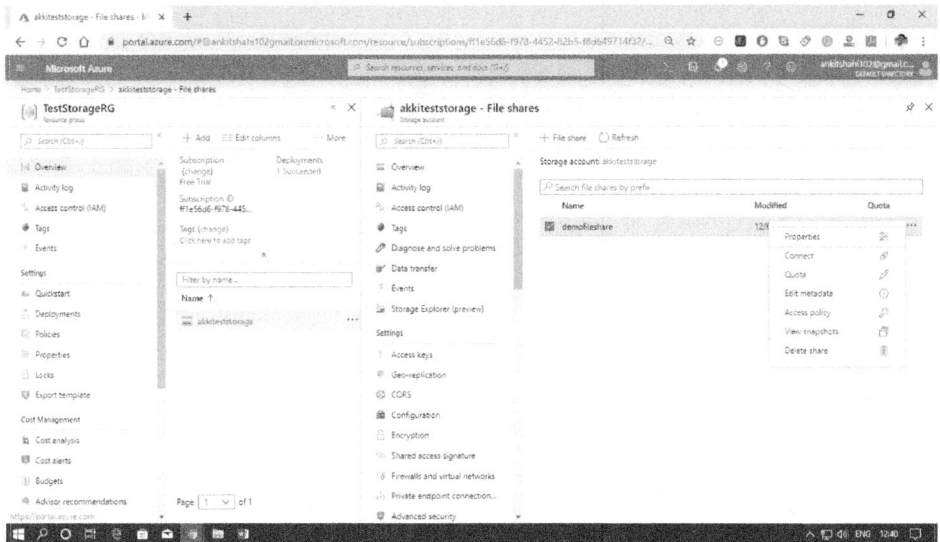

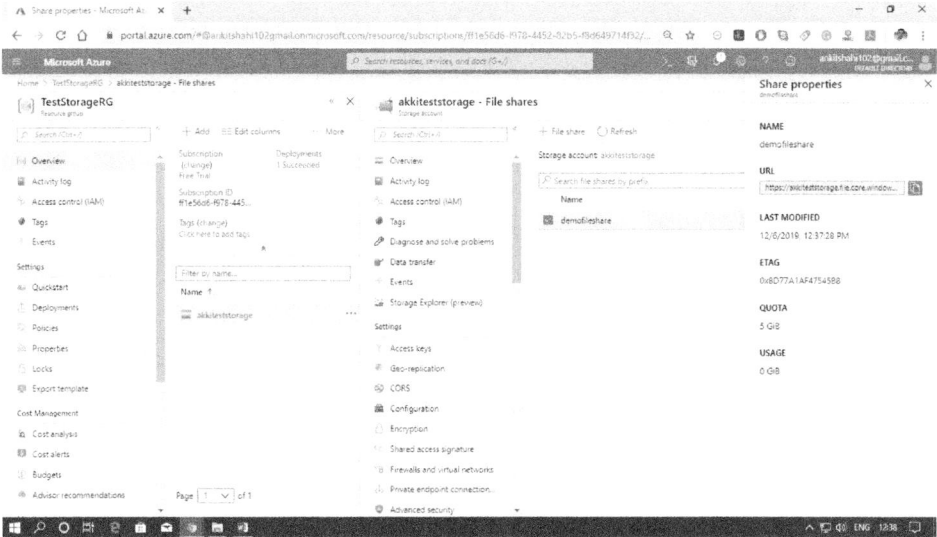

Step 5: And the second option is to connect, here you can see the PowerShell command and normal command to connect this File Share on a Windows computer.

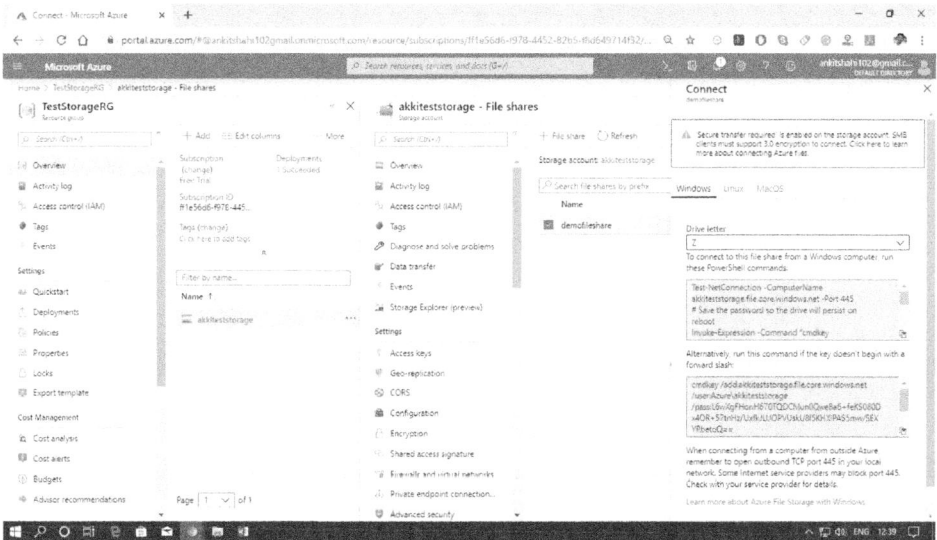

There are other options also given in the following figures.

Step 6: Click on File share and open it, where you can see the Access Control tab. You can use the Active Directory to control access to Azure file share. It is currently in preview.

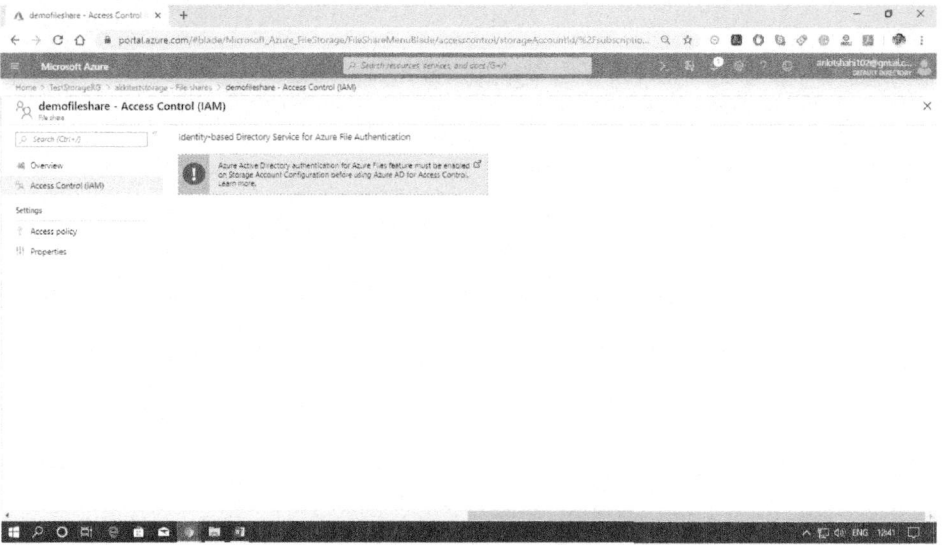

Step 7: To mount this file share with windows virtual machine Click on Connect.

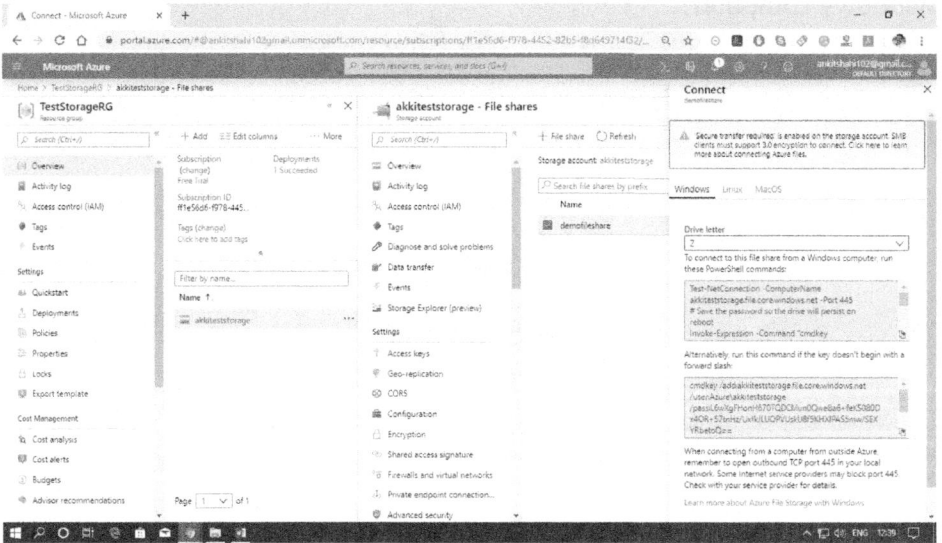

Step 8: Open the command line and copy the command given in the Connect window. If the command is executed successfully, your file share would be mounted with a virtual machine.

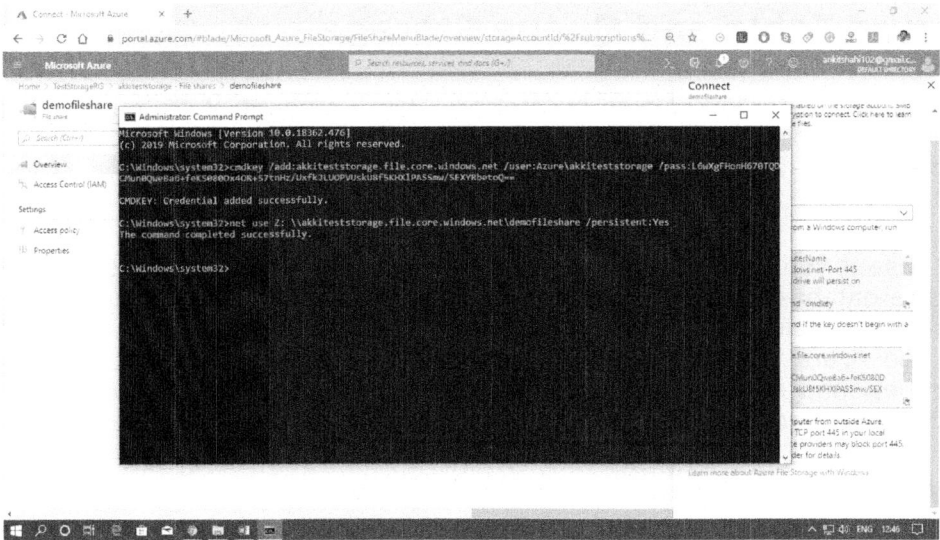

Azure Table and Queue Storage

Azure Table storage is used for storing a large amount of structured data. This service is a NoSQL data storage, which accepts authenticated calls from inside and outside of the Azure cloud. It is ideal for storing structured and non-relational data.

In case if you want to store relational data, then you should not use the Azure database. Unlike a relational database where the table has a fixed number of columns, and every row in the table should have those columns in the Azure table, which is a NoSQL data store. Each table can have entities, and each entity can have different properties. So generally, the schema will not be enforced on the objects that belong to a table.

Typical uses of Table storage include:

❖ Table storage is used for storing TBs of structured data capable of serving web-scale applications.
❖ It is used for storing datasets that don't require complex joins, foreign keys, or stored procedures and can be denormalized for fast access.
❖ It is used for quickly querying data using a clustered index.
❖ There are two ways of accessing data, one is using the OData protocol, and the other is LINQ queries with WCF Data Services with .NET Libraries.

Azure Table Structure

We need to create a storage account first because Azure table storage is offered under storage account, and then you have tables within that storage account. E.g., you can create employee table, address table, and each table will contain entities and entities will further include key-value pair like name email within an employee table.

However, one key difference here to the Azure table is the NoSQL data store and relational databases. These entities can have different schemas, so the first entity can have the name, email, and the second entity can have a name, email, and phone number also.

Azure Table structure

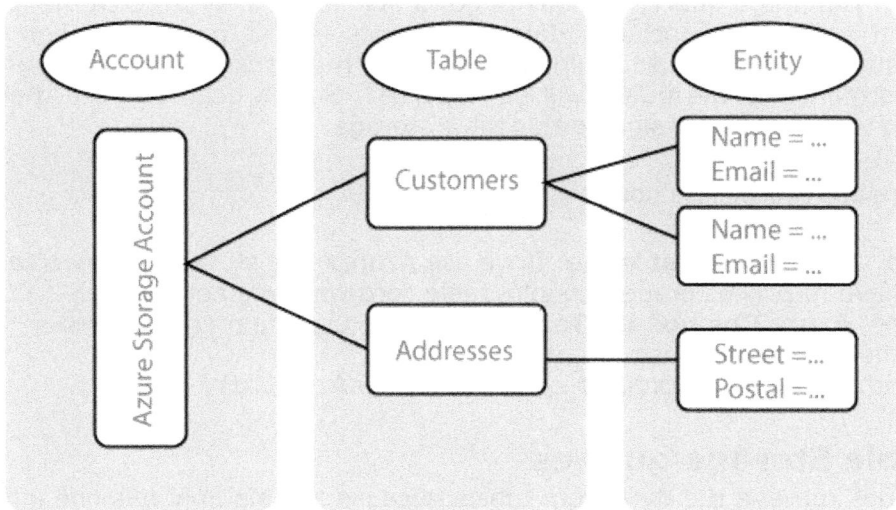

Azure table storage concepts

❖ **Accounts:** Every access to Azure storage service is done through a storage account, and all access to Azure Cosmos DB is done through a Table API account. So there are two types of tables Storage services available in Azure. The first one is Azure table storage, and the second one is a premium version, which is under Cosmos's DB. So if you are looking for a brilliant performance with low latency, then go for Cosmos's DB, particularly when you are dealing with mission-critical applications. In case if you can compromise on performance, but if you want to optimize the cost, then go for table storage.

❖ **Table:** It is a collection of entities. As we know, tables don't put a schema on entities that mean a single table can contain entities with different properties set.

❖ **Entity:** It is a set of properties, a similarly as database row. Azure Storage can be of 1MB in the size of the entity. But if we are using the premium version, which is Azure Cosmos DB, it can be of 2MB in size.

❖ **Properties:** It is a name-value pair where each entity can include up to 252 properties to store the data, and in addition to user properties, that means whatever the features you add. There are some system properties also that specify a partition key, a row key, and a timestamp. So every entity will have these three properties as a default. And when we are querying the data, we can carry the data based on the partition key and row key, and under the single partition, this row key should be unique. So when we query the data, we query the data with a partition key and row key. Generally, when we are fetching the entity from a single partition, it will be rapid because all

[67]

the objects belong to a separate partition will be stored in one server in the background within Azure.

When you find a query, it needs to go to one server only to fetch the data, but if your query includes data that exists in 2-3 partitions, then the question used to go to multiple servers in the background, thereby the performance of the query will be impacted. So we need to keep that in mind when we are designing this table storage.

There are two access points

> ❖ **Azure table storage:** If we use Azure table storage then we can have http://<storageaccount>.table.core.windows.net/<table>
> ❖ **Azure Cosmos DB Table API:** If we use the premium version then we need to use http://<storageaccount>.table.cosmosdb.Azure.com/<table>

Table Storage queries

Queries retrieve the data from tables because a table only has one index. Query performance is usually related to the PartitionKey and RowKey properties.

Here is a sample query to retrieve data from the server:

<account>.windows.core.net/registrations(PartitionKey="2011 NewYork City Marathon_Full", RowKey="1234_Ankit_M_55")

Azure Queue Storage Service

It is a queue service, but there is a more advanced version of the queue service that is available in Azure, which is a service bus queue.

> ❖ It is a service for storing a large number of messages in the cloud that can be accessed from anywhere in the world using HTTP and HTTPS.
> ❖ A queue contains a set of message. Queue name must be all lowercase.
> ❖ A single queue message can be up to 64KB in size. A message can remain in the queue for a maximum time of 7 days
> ❖ URL format is http://<storage account>.queue.core.windows.net/<queue>
> ❖ When the message is retrieved from the queue, it stays invisible for 30 seconds. A message needs to be explicitly deleted from the queue to avoid getting picked up by another application.

Azure Disk Storage

VM uses disks as a place to store an operating system, applications, and data in Azure. All virtual machines have at least two disks- a Windows operating system disk and a temporary disk. Both the operating system disk and the image are virtual hard disks (VHDs) stored in an Azure storage account. The VHDs used in Azure is .vhd files stored as page blobs in a standard or premium storage account in Azure. Virtual machines can also have one or more data disks that are also stored as VHDs.

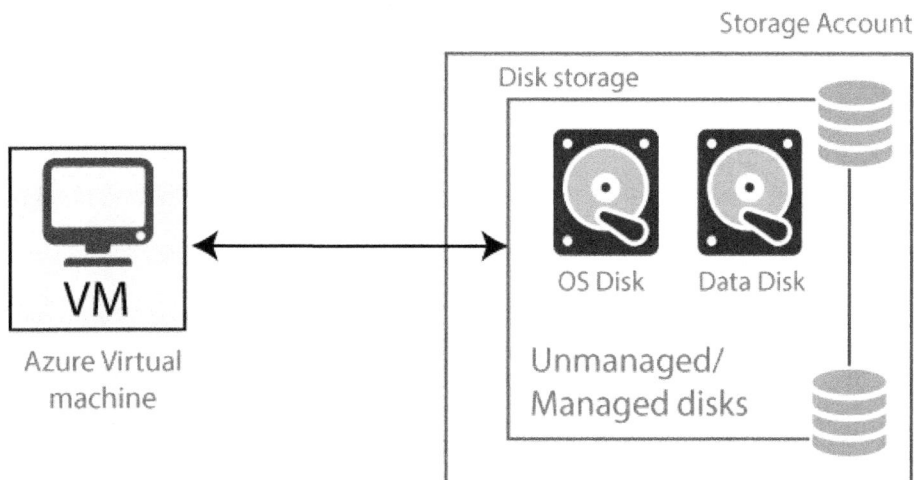

Storage Account

Disk storage

OS Disk   Data Disk

Unmanaged/
Managed disks

VM

Azure Virtual
machine

Temporary Disk: It is associated with the virtual machine that will be located in the underlying hardware from where the server is provisioned. So, the temporary disk will not be stored in a storage account. It will be stored in the underlying hardware from where this server is located.

Types of Disk

Different kinds of disks that are offered by Azure:

Unmanaged disks: It is a traditional type of disk that has been used by VMs. With these disks, we can create our storage account and specify that storage account when we create the disk. We must not put too many disks in the same storage account, resulting in the VMs being throttled.

Managed disks: It handles the storage account creation/management in the background for us and ensures that we do not have to worry about the scalability limits of the storage account. We specify the disk size and the performance tier (standard/premium), and Azure creates and manages the disk for us.

❖ **Standard HDD disks:** It delivers cost-effective storage. It can be replicated locally in one data-center, or be geo-redundant with primary and secondary data centers.

❖ **Standard SDD disks:** It is designed to address the same kind of workloads as standard HDD disks, but offer more consistent performance and reliability than HDD. It is suitable for applications like web servers that do not need high IOPS on disks.

❖ **Premium SSD disks:** It is backed by SSDs, and delivers high-performance, low-latency disk support for VMs running I/O-intensive workloads. The premium SSD disks are mainly used for production and database servers. So if we are hosting a database in a particular server, then the premium SSD will be a good option.

Microsoft recommends that we should use managed disks for all new VMs and convert our previous unmanaged to managed disks.

Disks backup

When we have this OS disk or data disk associated with Virtual Machine, we need to take the backup of the same regularly so that in case of data risk scenario, we can recover the data.

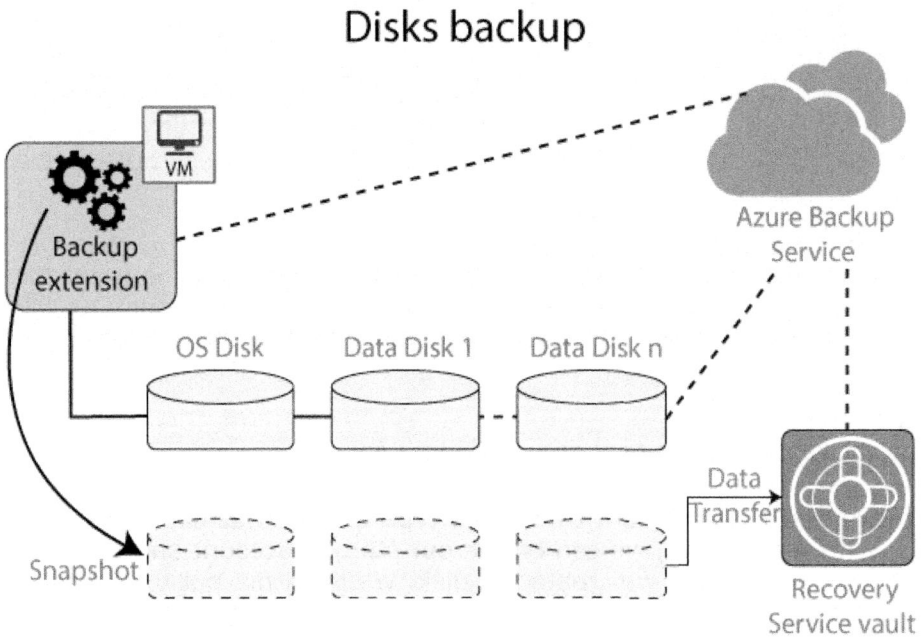

Disks backup

Azure provides the Azure backup service, which you can install as a backup extension on a particular VM and the extension based on the frequency you specified will take the snapshot off OS disk, and the data disk. And also, at different levels, so we can bring application-consistent snapshots, file consistent snapshots, and these snapshots will be moved into recovery service vault. That's where these snapshots will be stored. In case if something goes wrong with our VM or any particular data center is gone. We can still recover the virtual machine using these snapshots, and if we want to have a geo-redundant ability, then we can have this recovery services vault located in another region.

So for example, if our VM is located in northern Europe, then we can have a recovery service vault located in West Europe. In that way, we can protect our workloads against regional failure also.

Azure Storage Monitoring

Two capabilities are available in Azure for storage monitoring.

Continuous monitoring: Azure provides different metrics that are available both at the storage account level and individual service level also. These metrics are collected on an hourly basis, and we can define charts based on those metrics and pin those charts to the dashboard. We will see how to do that below.

Logging: We can enable client-side logging using Azure storage client library. And we can allow network logging, and server logging using Azure storage analytics. All these logging can be used to monitor an individual's transactions for continuous monitoring. These metrics are aggregated data, so we can't view an individual's transaction. But by enabling logging, we can investigate by going into the individual's transaction.

The essential tools that we use to monitor storage are audio storage analytics, which is explained below:

❖ Azure Storage Analytics performs logging and provides data quickly for the storage account. We can use this data to trace requests, analyze usages trends, and diagnose issues with our storage account.
❖ Metrics are enabled by default when we create a storage account. We can allow logging using the Azure portal, Rest APIs, or Client library. Metric uses the Get Blob Service properties, Get Queue Service Properties, Get Table Service Properties, and Get File Service Properties operations to enable Storage Analytics for all the services.
❖ The combined data is stored in a well-known blob (for logging) and in well-known tables (for metrics), which may use respective APIs service.
❖ Storage Analytics has a 20 TB limit on the amount of stored data that is independent of the total limit for your storage account.

Storage analytics logging:

Storage analytics records detailed information about successful and failed requests to a storage service. The data can be used to monitor individual requests and to diagnose issues with a storage service. Both authenticated and anonymous requests will be logged but at different levels. All logs are stored in block blobs inside a container named as $logs, which is automatically created when Storage Analytics is allowed for a storage account. The container ($logs) is located in the blob namespace of the storage account.

The logs are written in the following format

`<service-name>/YYYY/MM/DD/hhmm/<counter>.log`

Storage analytics metrics

Storage Analytics stores metrics, which include combined transaction statistics and capacity data about the request to a storage service. There are two types of storage analytics metrics.

Transaction metrics

❖ Transaction aggregated data recorded at hourly or minute like reading, write, update, etc.
❖ Data is recorded at the service level and API operation level

Capacity metrics

❖ Capacity data is recorded daily for a storage account's Blob service, which includes Capacity container count, object count, etc.

All the metrics data for each of the storage service is stored in three tables reserved for that service.

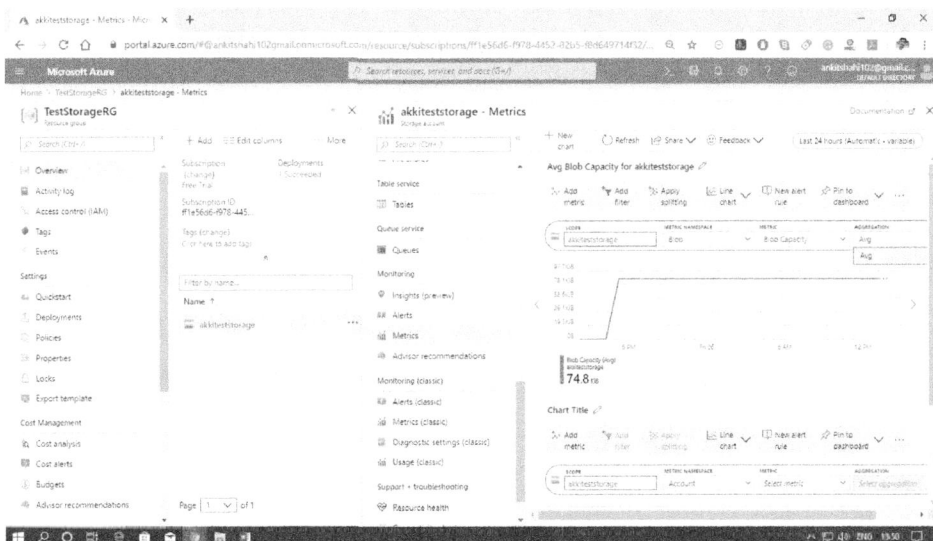

Azure Storage Resource Tool

Azure Storage Explorer: It is a standalone application that enables us to efficiently work with Azure Storage data on Windows, MacOS, and Linux. It provides several ways to connect to storage accounts. For example -

❖ We can connect to storage accounts associated with our azure subscriptions.
❖ We can connect to storage accounts and services that are shared from other Azure subscription
❖ We can connect and manage local storage by using the Azure Storage Emulator.

We can also connect to other services.

❖ Cosmos DB
❖ Data Lake store

Microsoft Azure Storage Emulator: It provides a local environment that emulates the Azure Blob, Queue, and Table services for development purposes. Using the storage emulator, we can test our application against the storage services locally, without creating an Azure subscription or incurring any costs. It is available as part of the Microsoft Azure SDK. We can also install the storage emulator by using the standalone installer.

It uses a local Microsoft SQL Server instance and the local file system to emulate Azure storage services. By default, the storage emulator uses a database in Microsoft SQL Server 2012 Express LocalDB.

Visual studio cloud & server explorer

Server explorer:

- The Azure Storage node in Server Explorer shows data in your local storage emulator account and your other Azure storage accounts.
- To see the storage emulator account's resources, expand the Development node.

❖ To view the resource in a storage account, expand the storage account's node in Server Explorer where you see Blobs, Queues, and Tables nodes.

Cloud explorer:

❖ Cloud Explorer enables us to view our Azure resources and resource groups. We can Inspect their properties, and perform key developer diagnostics actions from within Visual Studio.

To develop some solutions or applications based on Azure Storage resources, we can use **Azure storage client library**.

❖ We can use connection strings to connect to an Azure Storage account, then use the client libraries' classes and methods to work with blob, table, file, or queue storage.
❖ Install the NuGet package Windows Azure storage before start developing.

Management API's

❖ Create and manage Azure Storage accounts and connection keys with the management API.
❖ Install the NuGet package Microsoft.Azure.Management.Storage.Fluent.

Following are the steps to manage Azure storage resources using storage explorer.

Step 1: You have to download and install the storage explorer suitable for your OS.

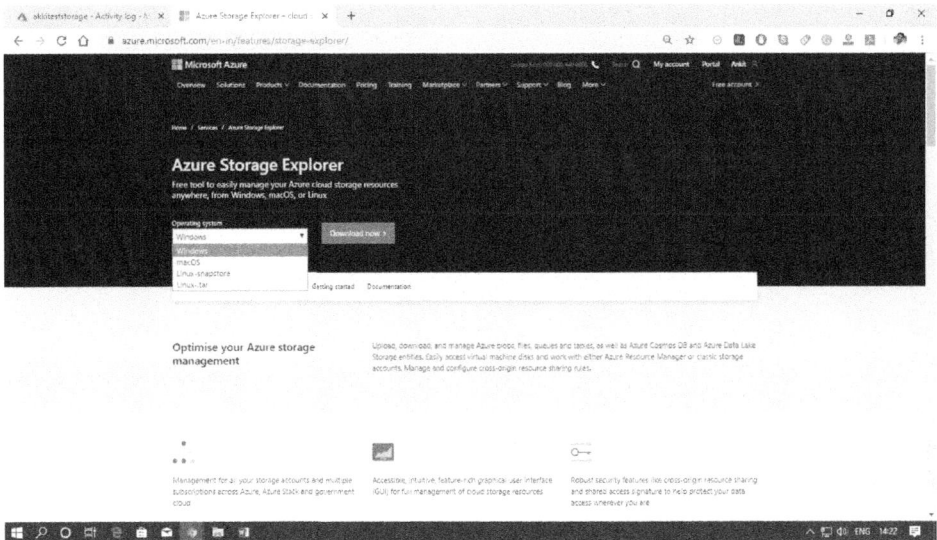

Step 2: When you install and open the storage explorer for the first time, the following screen will appear. Click next to continue.

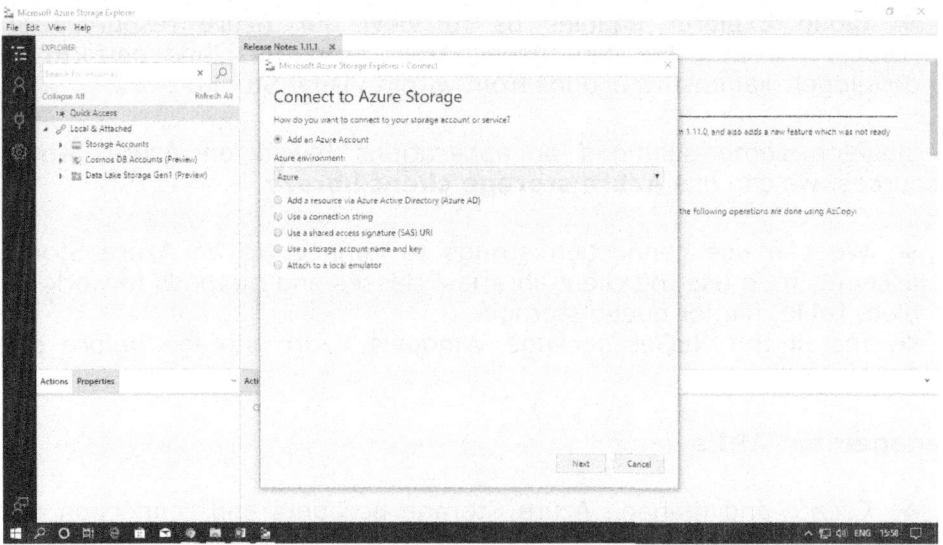

Step 3: Enter your login credentials to connect your Azure account with storage explorer.

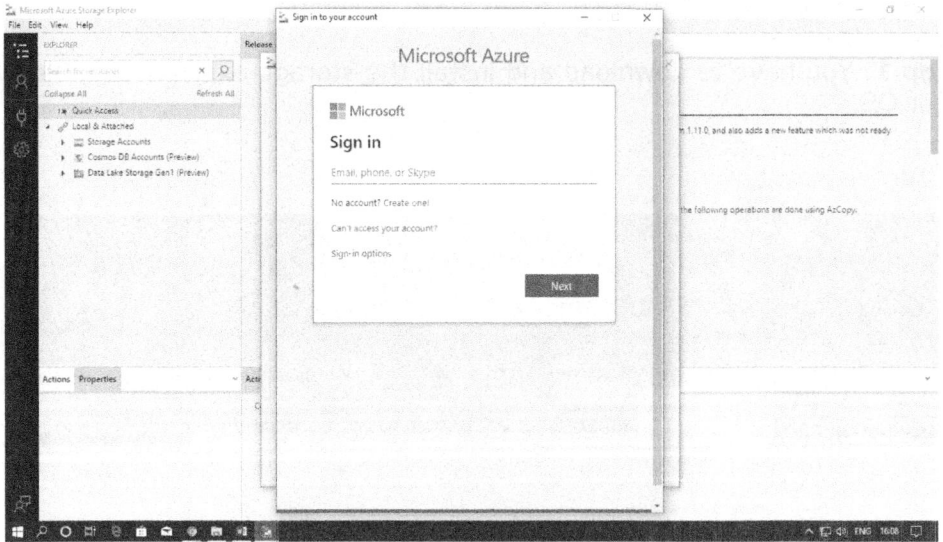

Step 4: Once you added the Azure account, you can select from which subscription you would like to view storage accounts. Then click on apply.

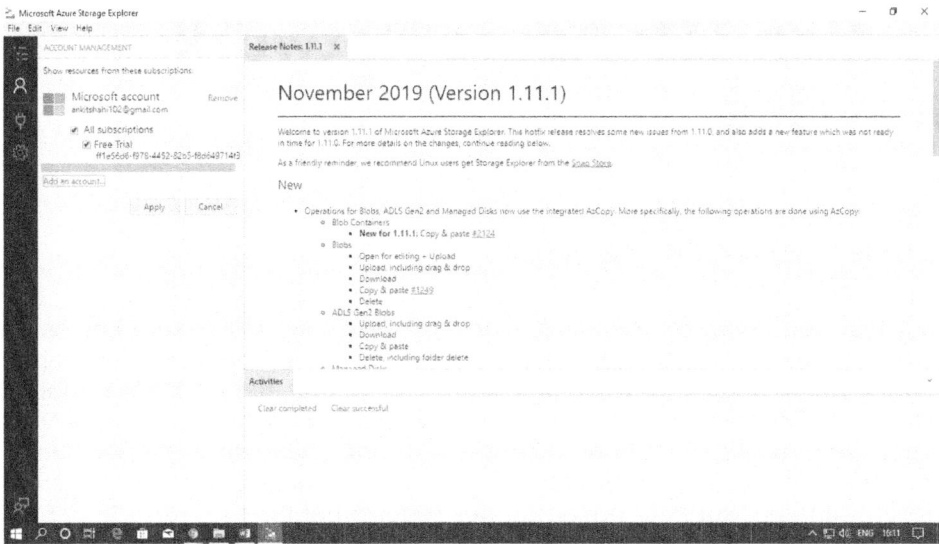

Step 5: You can see two nodes here, Local and attached, and the other is the selected storage account. You can see the containers, blobs, etc. here.

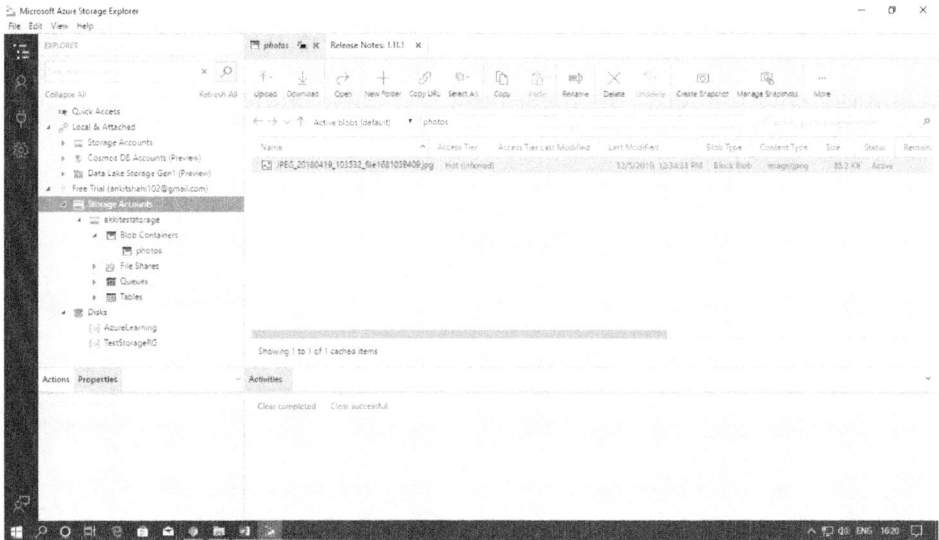

If you want to take some action, you can take steps to copy block containers, managed shared policies, set public access levels, acquire a lease, etc.

Section III: Azure Network Services

Azure Network Service

The most fundamental building block of Azure network services is the virtual network. Using a virtual network, we can deploy our isolated network on Azure. And we can divide the virtual network into multiple parts using subnets. For example - webserver subnet, App servers1 subnet, App servers2 subnet, Database subnet, Gateway subnet, Virtual Appliance subnet, etc. These are the typical examples, but we can create different kinds of subnets based on our requirements.

And once we create subnets, we can deploy different types of Azure services into these subnets. We can deploy a virtual machine into these subnets. But in addition to virtual machines, we can also deploy some specialized environments. i.e., some PaaS environments that are capable of being implemented into a virtual network. For example - in an app service environment, we can able to deploy in its own subnet. Similarly, there is something called **managed SQL** instance and also managed integration environment, all these kinds of environments we can able to deploy within a virtual network.

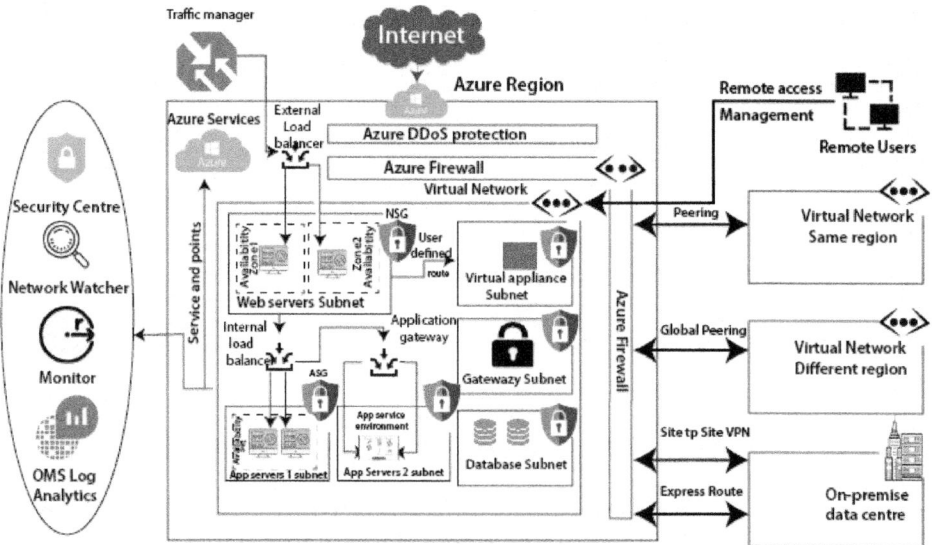

We can deploy different kinds of the appliance in a virtual appliance subnet like a firewall.

Service Protection: After the deployment of all these services, we need to protect these services. Azure provides several protection strategies.

DDoS Protection: The DDoS protection will protect our workload in the virtual network from DDoS attacks. There is a two-tier available in DDoS protection. One is the basic, which is free and enabled automatically. If we need the advance capability, then we can go for the DDoS standard tier.

Firewall: When we need network security, we use a firewall. Azure provides a firewall service which you can centrally manage inbound and outbound firewall rules. We can able to create network firewall rules, application firewall rules, inbound SNAT rules, outbound DNAT rules, etc.

Network Security Groups: If you think the firewall is too costly for you, then we can use Network security groups. We can filer the inbound and outbound traffic using network security groups. We can attach the network security group at two levels, one at the subnet level and other we can attach to a virtual machine.

Application Security Groups: Microsoft introduces the application security group to put all the server related to one application in one application security group and use that application security group in network security group inbound and outbound rules. The primary purpose of the Application Security Group is to simplify the rule creation in NSG's.

Service Availability

We have to make sure that our application is highly available and resilient to regional failures, data center failure, and rack failures. Azure provides some services to make our application highly available; these are:

Traffic Manager: Microsoft Azure traffic manager controls the distribution of user traffic for service endpoints in different regions. Service endpoints supported by Traffic Manager include Azure VMs, Web Apps, Cloud services, etc. It uses DNS to direct the client request to the right endpoint based on a traffic-routing method and the health of endpoints.

Load Balancer: Load balancer is used to distribute the traffic evenly between a pool of web servers or application servers. There are two types of the load balancer, one is external load balancer which sits outside the virtual network and the second one is an internal load balancer that sits inside the virtual network.

Application Gateway: Using the application gateway, we can achieve URL path-based routing, Multi-site hosting, etc.

Availability Zones: By deploying our virtual machines into different availability zones, we can route our application traffic to virtual machines that are located in different availability zone in case of failure of datacenter within any region.

Communication

The basic idea behind creating a virtual network is to enable communication between workloads using default system routes. These system routes will be deployed by Azure automatically. But we can also override these system routes and configure our user-defined routes; then, we can do that too.

Peering: To enable communication between two virtual networks, we can establish peering. We can do this peering with virtual networks within the same region. If we have an Azure virtual network in another region, then we can use global peering. And for the on-premises data center, we have two options, and one is the site to site VPN, which will get established over the Internet. But for private connectivity, we have to use the express route.

Monitoring: Once we deployed all the services from the networking perspective, we need to start monitoring them. Azure provides some services to monitor traffic.

Security Center: It is a central security monitoring tool using which we can view the Security score of your overall deployment, and any recommendation generated by Azure based on the security policies we have applied. Both from networking and also the service deployed on that virtual network.

Azure Virtual Network

The Azure Virtual Network is a logical representation of the network in the cloud. So, by creating an Azure Virtual Network, we can define our private IP address range on Azure, and also deploy different kinds of Azure resources. For Example - Azure virtual machine, App service environment, Integration service environment, etc.

Azure Vnet Capabilities

Following are the capabilities of the Azure Vnet:

Isolation and segmentation: To deploy resources such as virtual machines into virtual networks, they will be isolated from other resources. By putting the virtual machine into your virtual network, it cannot be reached from the Internet or other Azure resources unless we enable communication in between. We can also use subnets within virtual networks to further segment our resources within the network.

Communication with the Internet: All resources in a virtual network can communicate outbound to the Internet by default. But it needs to establish an inbound connection from the Internet. We can either use public IP or load balancers.

Communication between resources: Communication between the number of resources inside the virtual network or with other resources through service endpoints.

Communication with on-premises resources: By establishing either point to site VPN or site to site VPN or Express route, your workloads within Azure virtual network can seamlessly communicate with workloads within our on-premises data center.

There are lots of capabilities within the Azure virtual network that we can use to control the traffic.

Filter network traffic: We can use Network Security Groups, Application Security Group, Azure firewall, or third-party network virtual appliance to filter the traffic coming to the resources in the virtual network.

Route network traffic: We can route the network traffic using the routing tables, we can configure user-defined routes to route all the outbound traffic, let's say via a firewall.

Monitor network traffic: By network security groups and traffic analytics monitoring solution, you'll be able to carry out extensive monitoring on both inbound and outbound communications.

Subnet

Subnet plays a vital role because many configurations will be done at a subnet level. It is a range of IP addresses in the VNet. Vnet can be divided into multiple subnets based on different design considerations, for example - we can deploy a virtual machine, App services environment, integration service environment, etc. VMs & PaaS services deployed to subnets n the same VNet and can communicate with each other without any extra configuration. Route tables, NSG, Service endpoints, and policies are configured to the subnets.

Creating Azure Virtual Network and subnets

Step 1: Select your existing resource group, or you can create a new resource group.

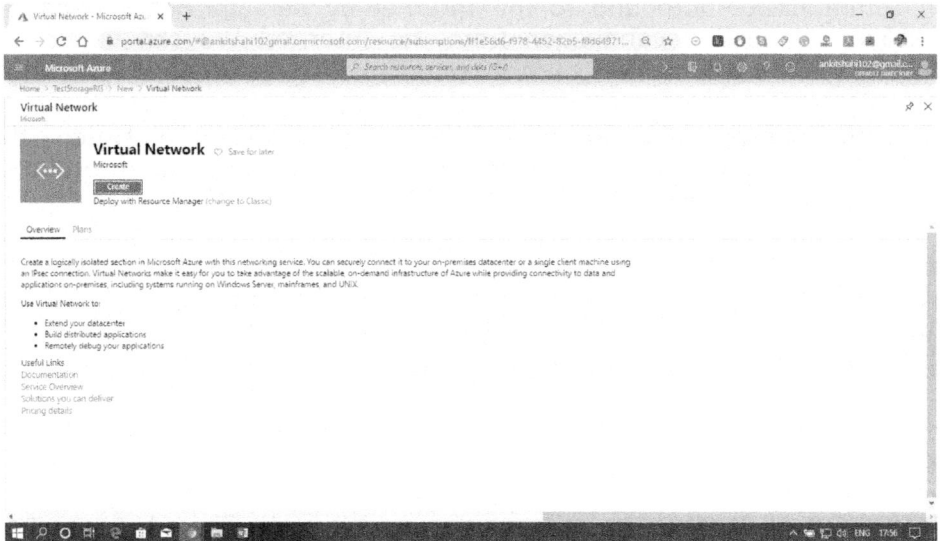

Step 2: After opening your resource group, click on Add then type in Virtual network in the search box. Click on Create.

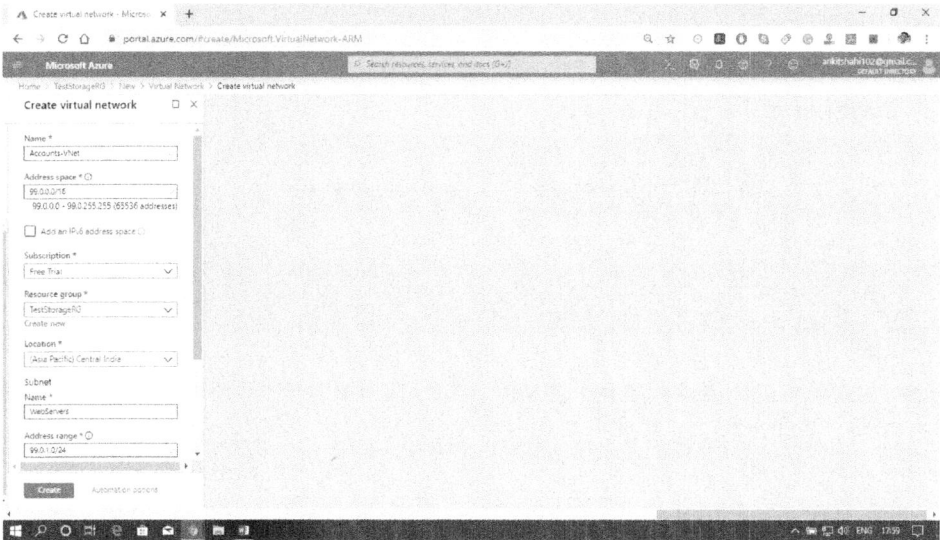

Step 3: A new window will appear, where you need to fill the details like - name, address space (e.g., 99.0.0.0/16), Name of the subnet, subnet address space (e.g., 99.0.1.0/24). Leave everything as it is and click on create.

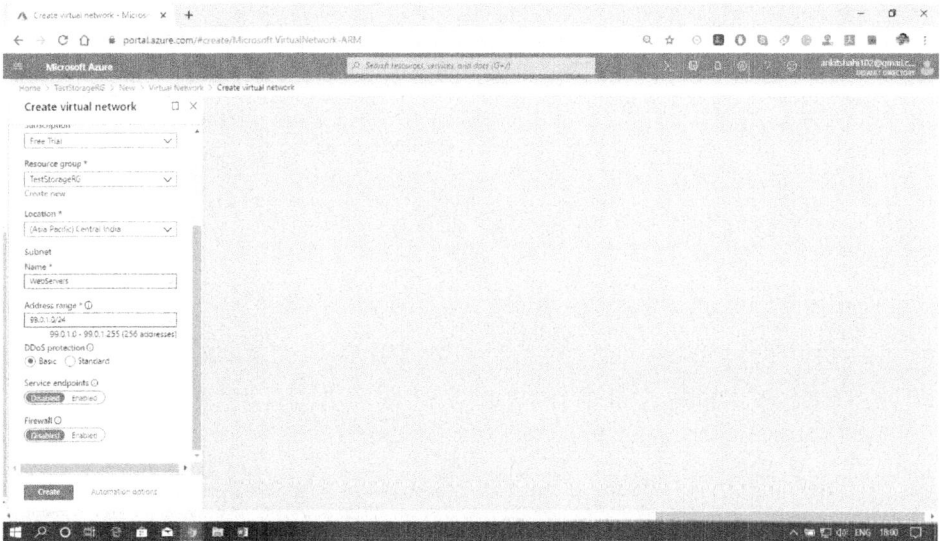

Step 4: Now, your Vnet is created. Let's add a subnet into it. Click on the subnet, then click on add subnet.

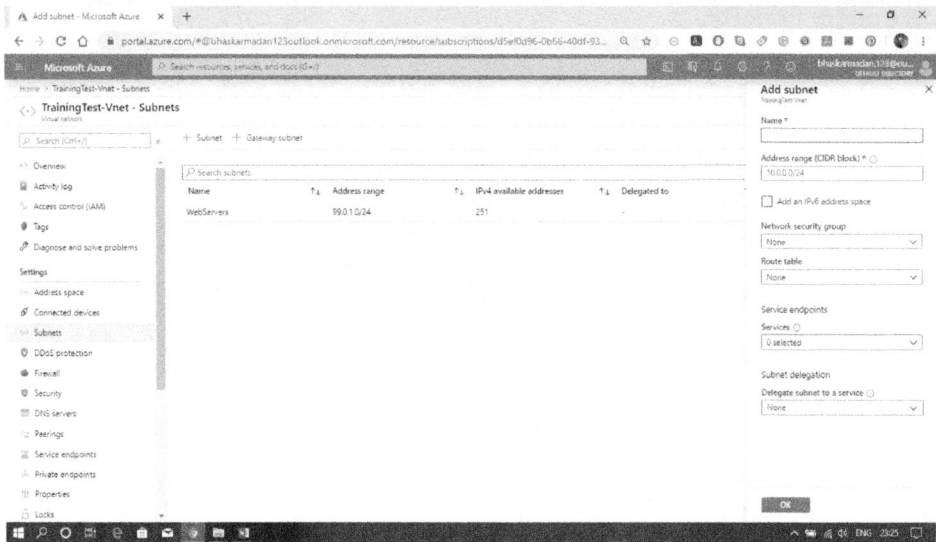

Step 5: On the next slider window, give a name to the subnet you want to create, provide the address range (if the address range is currently in use, you cannot change it). Then click on the ok button to create the subnet.

Azure Network Security

A network security group consists of security rules that allow or deny inbound/Outbound network traffic to or from different types of Azure resources that we will host in our Azure virtual network. And we can apply the network security group at different levels. For example:

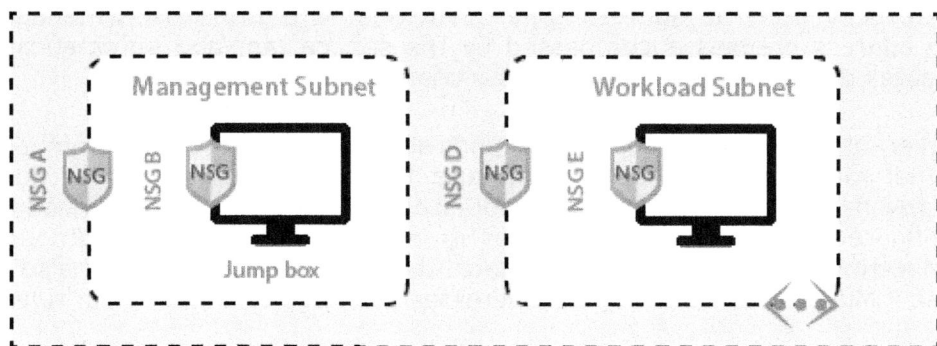

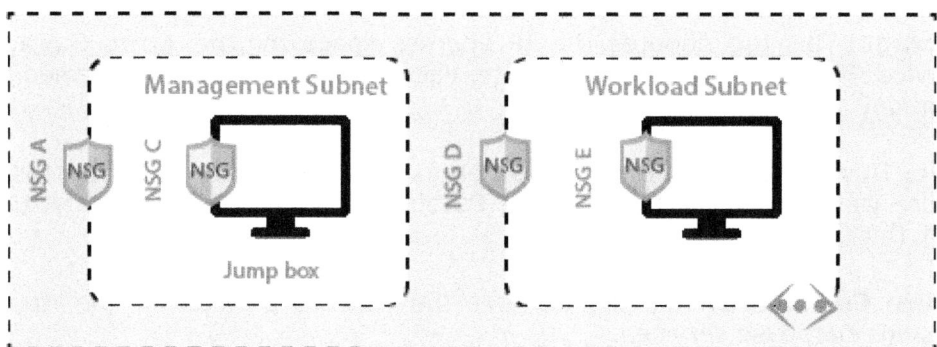

Security rule properties:

Name: The name of the network should be unique within the network security group.

Priority: Security rules are processed in priority order with a lower number has the highest priority.

Source or Destination: (The IP address, CIDR (Classless inter-domain routing) block, service tag, or application security group) The ability to specify multiple individual IP addresses and ranges in a rule is referred to as augmented security rules.

Protocol: TCP, UDP, etc.

Port range: we can specify an individual or range of ports

Action: Allow or Deny

Service Tags

Service tag represents a group of IP address prefixes to help minimize complexity for security rule creation. We cannot create our service tag, nor specify which Ip address is included within a tag. Microsoft manages the address prefixes encompassed by the service tag, and automatically updates the service tag as an address change.

Earlier, if we want to allow communication to Azure service from our virtual machine, we need to configure IoT of outbound rules because Microsoft is providing list of IP addresses for each service you need to configure those list of IP addresses in our NSG rule to allow outbound connection from our virtual machine to that particular service and also in case if Microsoft is changing the addresses you need to change your rules.

Using service tags will simplify your NSG rules a lot, for example:

Storage: This tag denotes the IP address space for the Azure Storage service. If you specify Storage for the value, traffic is allowed or denied to storage.

SQL: This tag denotes the address prefixes of the Azure SQL Database, Azure Database for MySQL, Azure Database for PostgreSQL, and Azure SQL Data Warehouse services.

Azure CosmosDB: This tag denotes the address prefixes of the Azure Cosmos Database services.

AzureKeyVault: This tag denotes the address prefixes of the Azure KeyVault service. If you specify AzureKeyVault for the value, traffic is allowed or denied to AzureKeyVault.

EventHub: This tag denotes the address prefixes of the Azure EventHub service. If you specify EventHub for the value, traffic is allowed or denied to EventHub.

Default Rules

Some default rules are created by default when we create NSG. There are two types of default rules.

Inbound Security rules

❖ **AllowVNetInbound:** Traffic is allowed from any resources within the VNet

❖ **AllowAzureLoadBalancerInbound:** Any traffic originating from Azure load-balancer to any of the virtual machines within the network is permitted.

❖ **DenyAllInbound:** By default, virtual machines in the virtual network can communicate with each other, and also Azure load balancer can communicate with virtual machines within the virtual network.

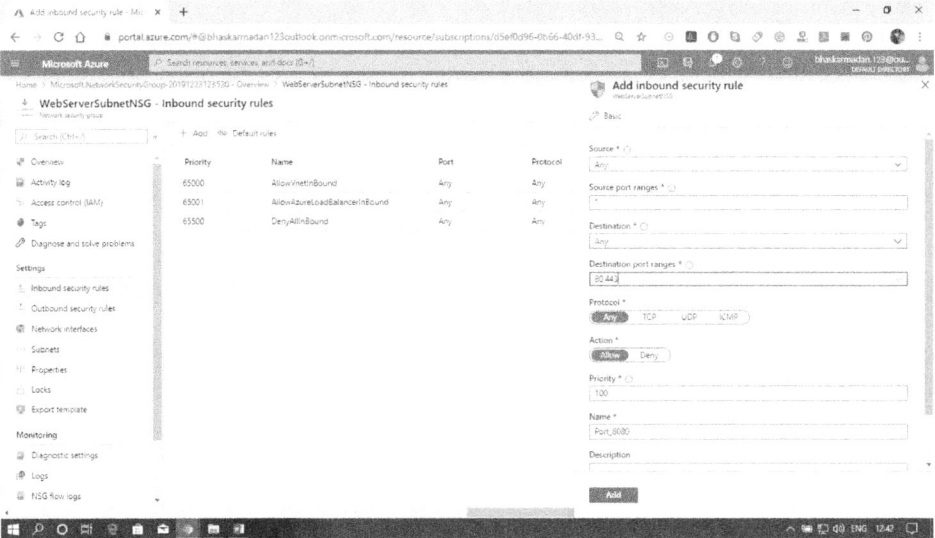

Outbound Security rules

❖ **AllowVNetOutBound:** Traffic is allowed through any resources within the VNet

❖ **AllowInternetOutBound:** Traffic originating from any resources in the VNet to the Internet is allowed.

❖ **DenyAllOutBound:** By default, virtual machines in a virtual network can communicate with each other, and also Azure load balancer can interact with the virtual machine within the virtual network.

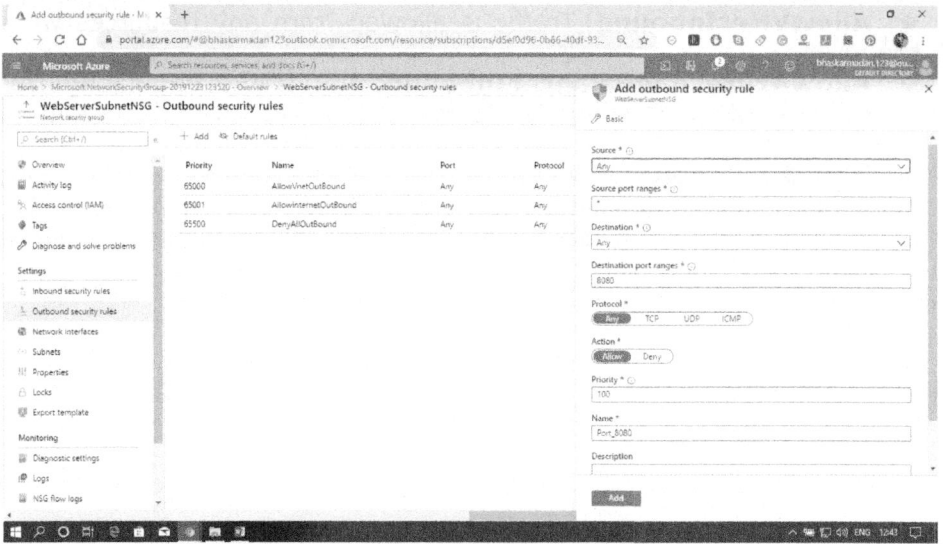

Application Security Groups

Application security groups enable you to configure network security as a natural extension of an application's structure, allowing you to group virtual machines and define network security policies based on those groups. For example -

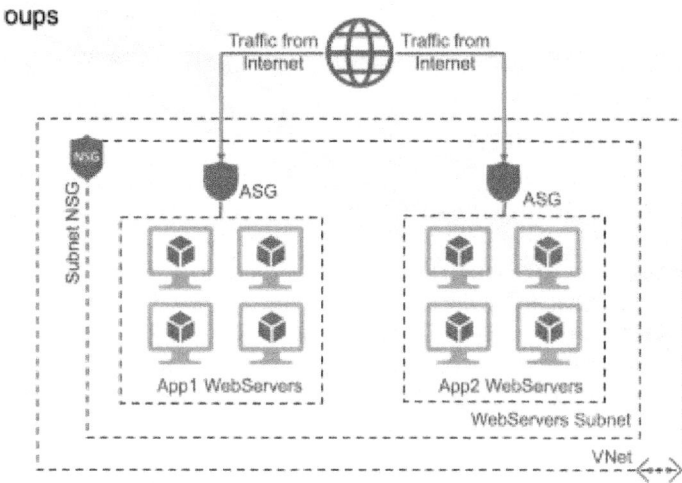

Configuring an NSG at Subnet and VM level

Step 1: Click on create a resource button and type-in Network Security Group. Then select Network Security Group, and click on create button.

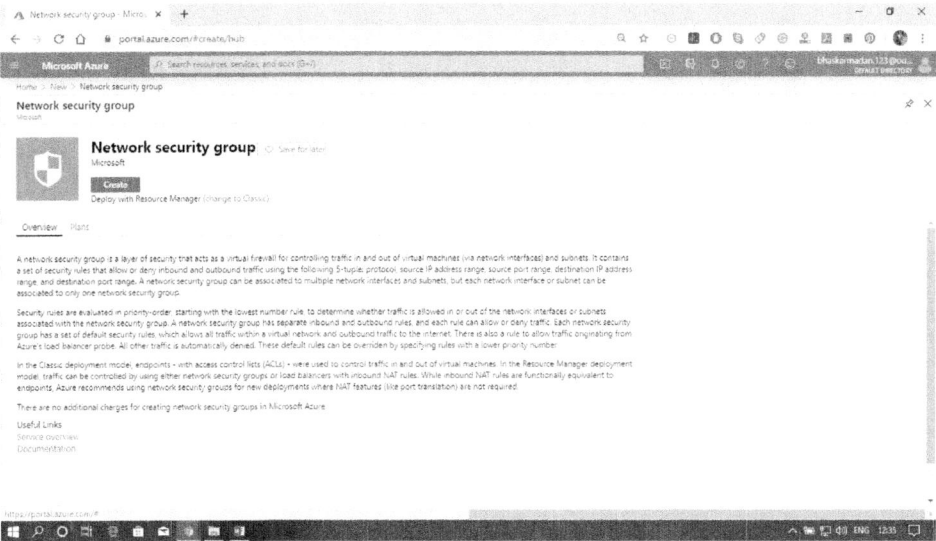

Step 2: Now, you are on the Network Security Group creation page. Select the resource group, fill the name, select the region, and click on review+create.

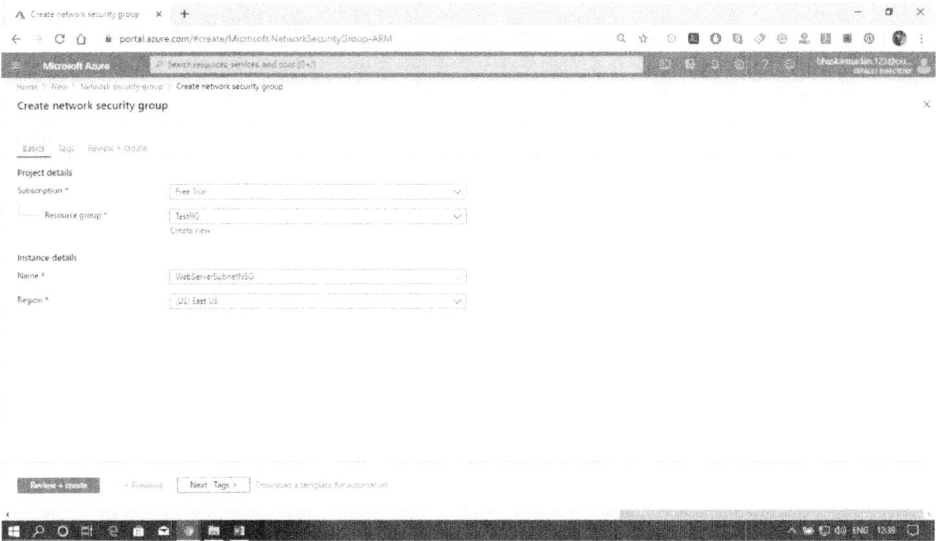

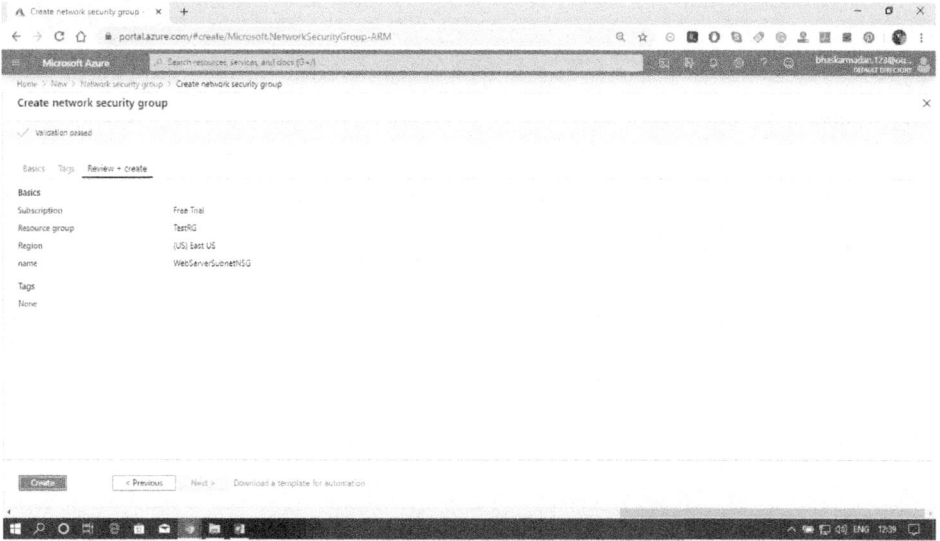

Step 3: Your NSG is created, now we will associate this NSG with the subnet.

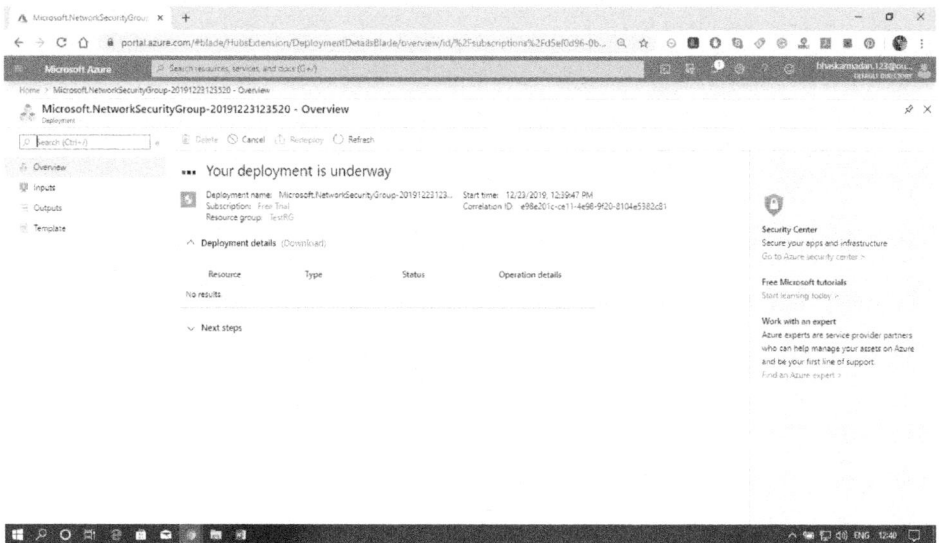

Step 4: Click on the subnet, then click on add Associate. Select the virtual network and subnet with which you want to associate this NSG.

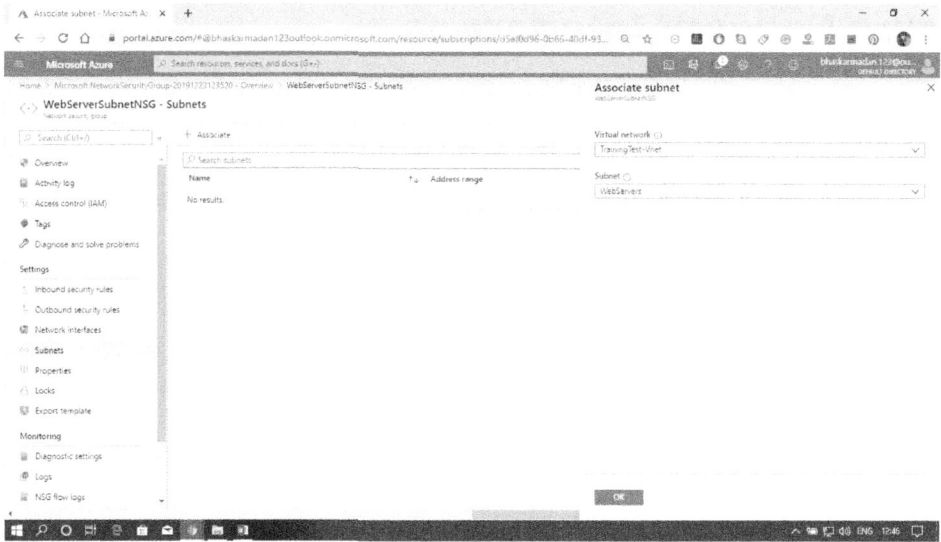

Step 5: Finally, click on the ok button. Your NSG is now associated with the subnet.

Azure Network Interface

A Network Interface (NIC) is an interconnection between a Virtual Machine and the underlying software network. An Azure Virtual Machine (VM) has one or more network interfaces (NIC) attached to it. Any NIC can have one or more static or dynamic public and private IP addresses assigned to it.

Configuring the network interface

Virtual network & subnets: we can attach a network interface to a VNet and Subnet, and once we deployed a NIC into a VNet, we can't change it.

IP configuration: Public and private IP addresses will be assigned at the NIC level. Primary & secondary IP configurations

NSG & Routes: We can apply zero or one network security group and one or more routes to a network interface.

IP Forwarding: This setting must be enabled for every network interface that is attached to the virtual machine.

DNS servers- We can specify which DNS server the Azure DHCP servers assign a network interface.

IP addresses
It is a unique reference that identifies each computer using the Internet Protocol to communicate over a network.

There are two ways to define an IP address:

Private IP Addressing	Public IP addressing
Private IPv4 addresses enable a virtual machine to communicate with other resources in a virtual network or other connected networks.	When we assign a public IP address to an Azure resource that supports public IP addresses, which enables Inbound communication from the internet to the resource, resources like Azure VMs and Outbound connectivity to the internet using a predictable IP address is called a public IP address.

When we select the dynamic addresses, Azure automatically assigns the next available address from the address space of the subnet you selected.	The dynamic addresses are released when a public IP address resource will be dissociated from a resource it is associated to.
When we select static addresses, we must manually assign an available IP address from within the address space of the subnet you selected.	Static IP addresses are assigned to the machine when a public IP address is created.

Hostname resolution

We can specify a DNS domain name label for a public IP resource, which creates a mapping for domainnamelabel.location.cloudapp.azure.com to the public IP address in the Azure-managed DNS servers.

Internal DNS hostname resolution (for VMs)

All Azure VM's are configured with Azure-managed DNS servers by default unless you want to set custom DNS servers explicitly, and these Azure managed DNS servers provide internal name resolution for VM's that reside with the same VNet. So if we want to RDP from one virtual machine to another virtual machine, you actually can use the name of the machine in stuff Private IP address.

Configure multiple NICs and IP addresses for a VM

Step 1: Click on Create resource button and type-in network interface. Then click on Network Interface and create.

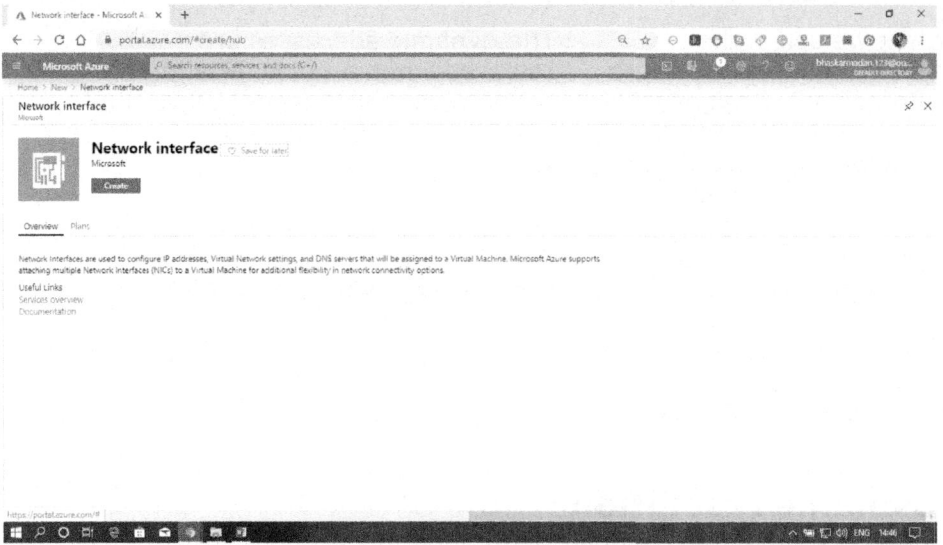

Step 2: Now, fill the required details and click on create.

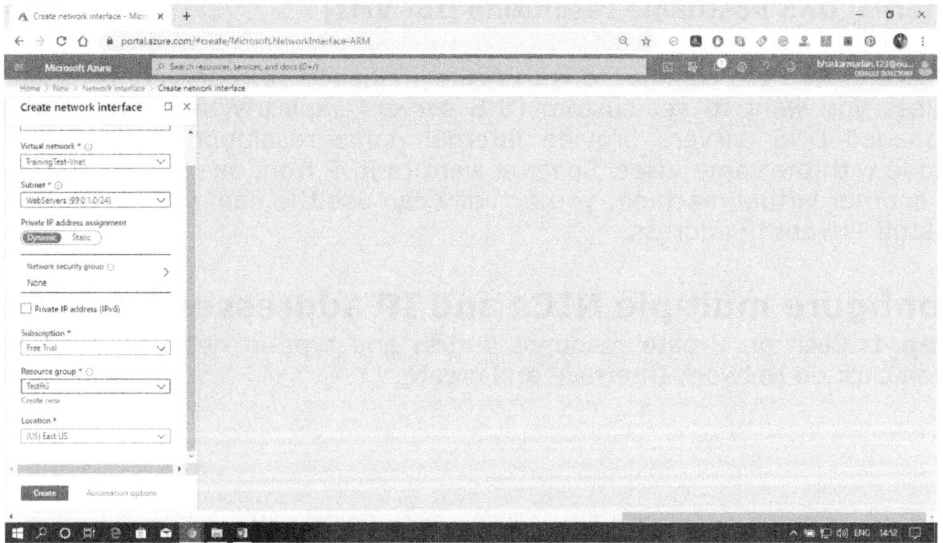

Step 3: Your Network Interface will be created and ready to embed.

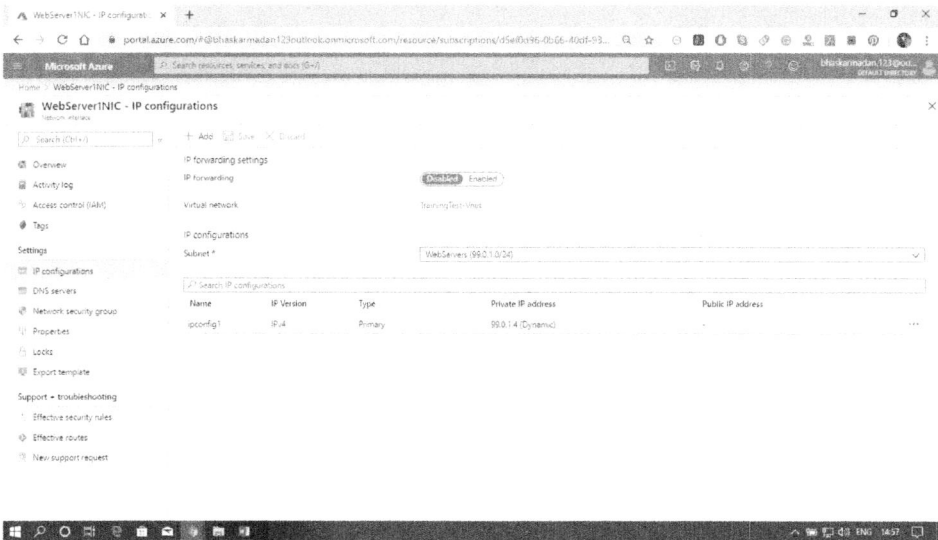

Step 4: Again, go to the home page and create a public IP address.

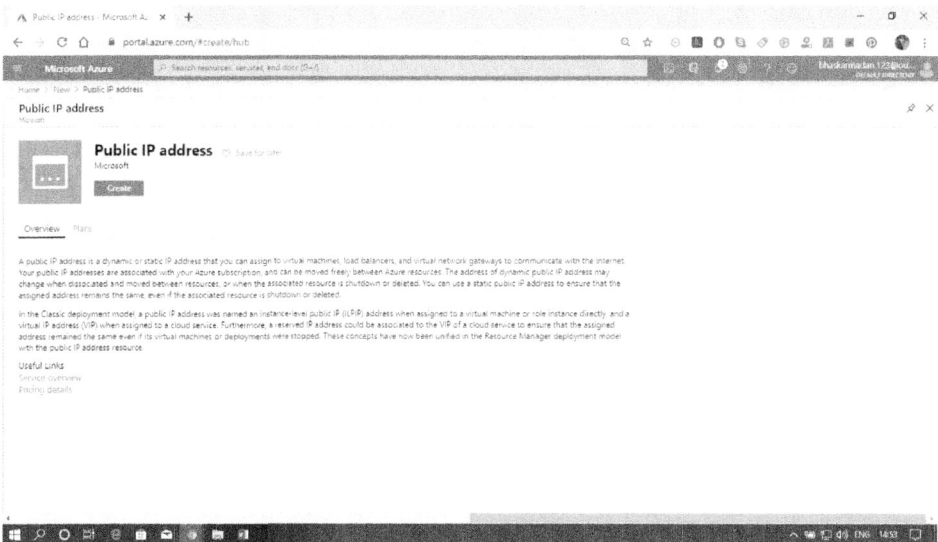

Step 5: Now, fill the required details and click on create.

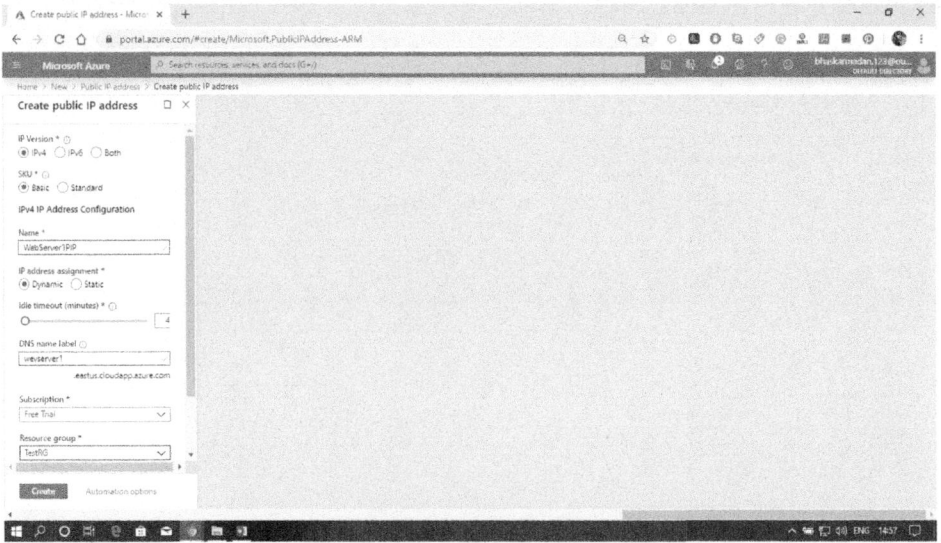

Step 6: You have now both the NIC and IP address ready to use with the virtual machine.

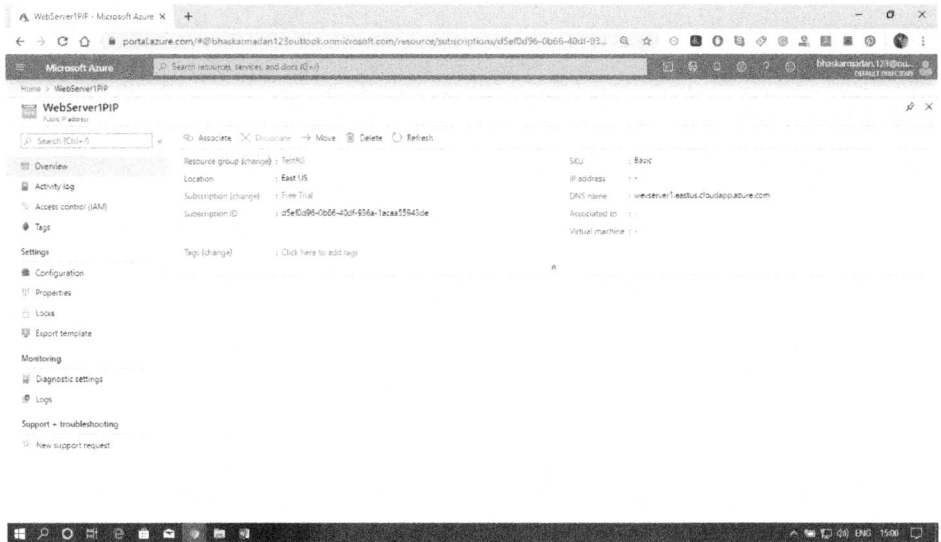

Azure Availability Zones and Sets

If we want to migrate a mission-critical application into Azure and because of the nature of the application, one of the key requirements is to make an application resilient to both reasonable failures, data center failure, and also even rack failure. So, to make that application highly available in all circumstances, we use Azure services that can deliver that requirement in terms of making application resilient to reasonable failures.

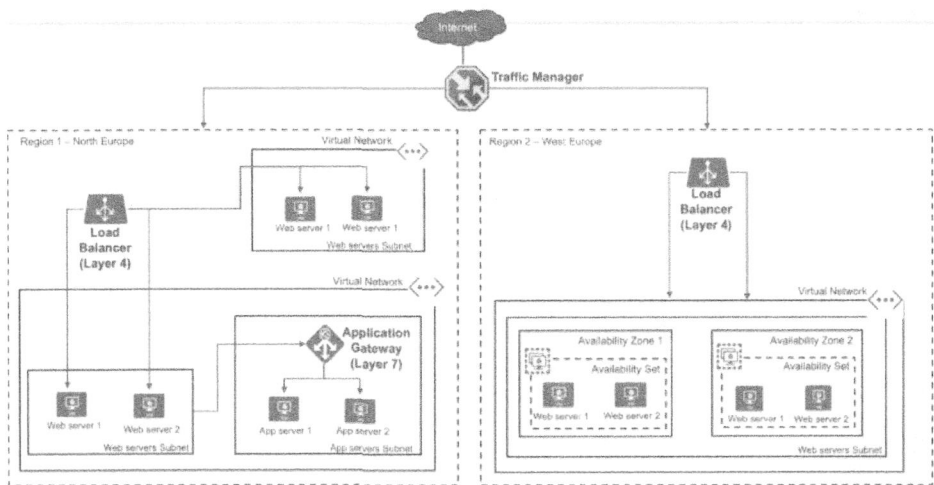

Traffic Manager: We can use traffic manager to monitor the endpoints located in different regions, and if any of the endpoints are no longer reachable, then all the traffic can be routed to other regional endpoints. It contains different routing methods, such as priority weightage, performance, geographic, etc.

Azure Load balancer: We can use Azure load balancer to balance the traffic between our web servers or application servers. It offers layer-4 load balancing, i.e., if we use source IP, source port, destination IP, destination port, and protocol using which we can configure the rules in load balancer to load balance traffic between a form of web servers.

Application Gateway: If we want to load-balance traffic based on URL based routing, or we can say that we want to host multiple sites on the same public IP address and other things, then we can use the application gateway.

To deliver high availability, Azure provides two more important features.

Availability Zone: It is a high availability offering that protects your application and data from data center failures. Generally, every Azure region consists of multiple data centers located at different physical locations. When you are deploying your services into Azure, you can able to select into which availability zone you want to deploy your service.

Availability Set: It works at a rack level. It is a logical grouping of the virtual machine within the data center that allows Azure to understand how your application is built to provide for redundancy and availability. The availability set consists of two domains one is fault domain, and another is the update domain.

❖ **Fault Domain:** It is a logical group of the underlying hardware that share a common power source and network switch similar to a rack within an on-premises data center. So, if we are deploying all the virtual machines into the same fault domain, then any hardware failure will knock out all the virtual machines in that particular rack or fault domain.

❖ **Update Domain:** It is a logical group of the underlying hardware that can undergo maintenance or be rebooted at the same time because Azure will do infrastructure management.

❖ **Managed Disk fault domains:** For VMs using Azure Managed Disks, VMs are aligned with managed disk fault domains when using a managed availability set. This alignment ensures that all the managed disks attached to a VM are with the same managed disk fault domain.

Azure Load Balancer

The load balancer is used to distribute the incoming traffic to the pool of virtual machines. It stops routing the traffic to a failed virtual machine in the pool. In this way, we can make our application resilient to any software or hardware failures in that pool of virtual machines.

Features of Azure load balancer

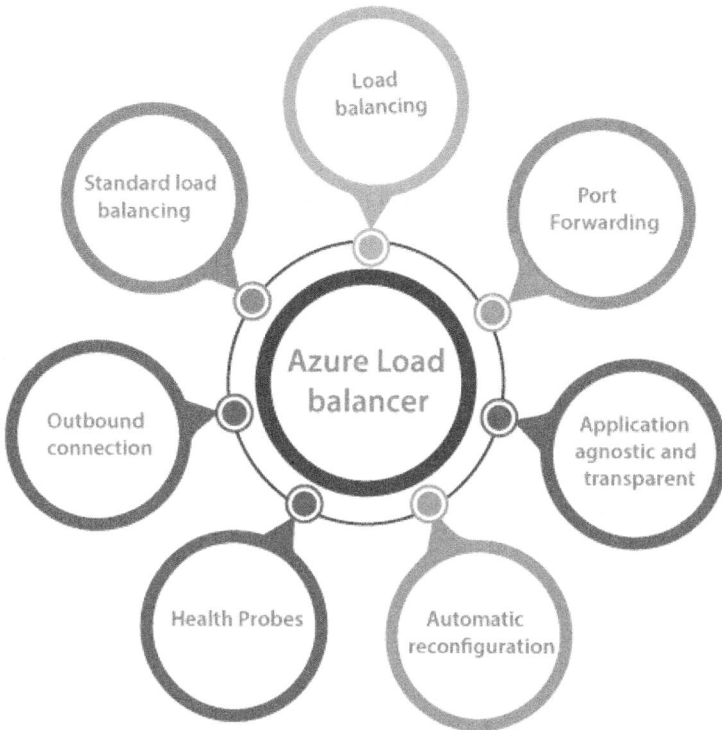

❖ **Load Balancing:** Azure load balancer uses a 5-tuple hash composed of source IP, source port, destination IP, destination port, and protocol. We can configure a load balancing role within the load balancer in such a way based on the source port and source IP address from where the traffic is originating.

❖ **Port forwarding:** Load balancer also has port forwarding capability if we have a pool of web servers, and we don't want to associate public IP address for each web server in that pool. If we're going to carry out any maintenance activities, you need to RDP into those Web servers having a public IP address on that web servers.

- ❖ **Application agnostic and transparent:** Load balancer doesn't directly interact with TCP or UDP or the application layer. We can route the traffic based on URL or multi-site hosting, and then we can go for the application gateway.
- ❖ **Automatic reconfiguration:** Load balancer can reconfigure itself when we scale up or down instances. So, if we are adding more virtual machines into the backend pool, automatically load balancer will reconfigure.
- ❖ **Health probes:** As we discussed earlier, the load balancer can recognize any failed virtual machines in the backend pool and stop routing the traffic to that particular failed virtual machine. It will recognize using health probes we can configure a health probe to determine the health of the instances in the backend pool.
- ❖ **Outbound connection:** All the outbound flows from a private IP address inside our virtual network to public IP addresses on the Internet can be translated to a frontend IP of the load balancer.

Configuration elements of Load Balancer

- ❖ **Front-end IP configuration:** It is the IP address to which the incoming traffic will initially come to, and Azure load balancer can have one or more front end IP addresses. They are sometimes also called as virtual IPs.
- ❖ **Back-end address pool:** These are the pool of virtual machines to which the traffic will eventually go to.
- ❖ **Load balancing rules:** A load balancing rule is simply a mapping between the front end IP configuration and back-end address pool.
- ❖ **Probes:** Probes enable us to keep track of the health of VM instances. If a health probe files, the VM instance will be taken out of rotation automatically.
- ❖ **Inbound & Outbound NAT rules:** NAT rules defining the inbound traffic flowing through the front end IP and distributes to the backend IP. Outbound rules will transmit VM private IP to load balancer public IP.

Creating Azure Load Balancer

Step1: Go to the Azure portal, and click on create a Resource. After that, type-in Load Balancer, and click on it.

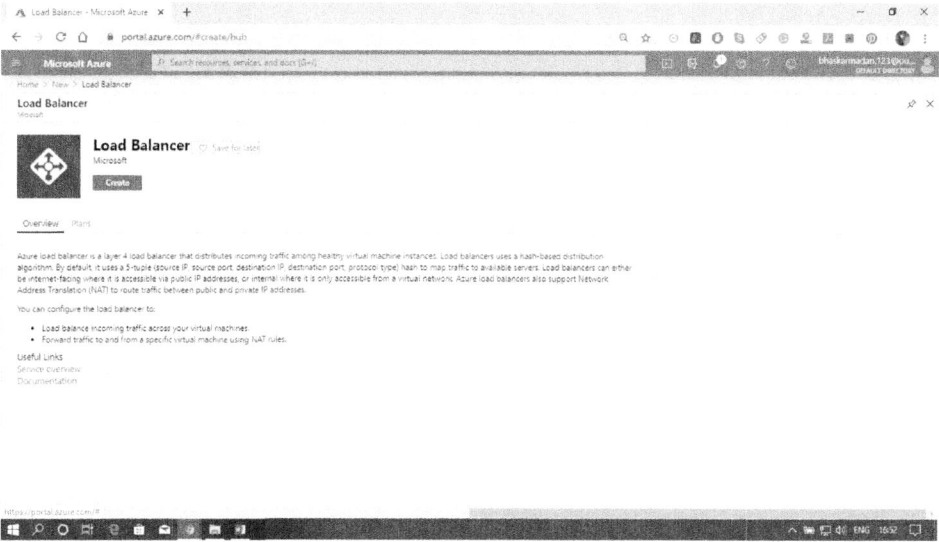

Step 2: You are now on the Load Balancer creation page. Fill all the required details as the figure below. And click on review+create.

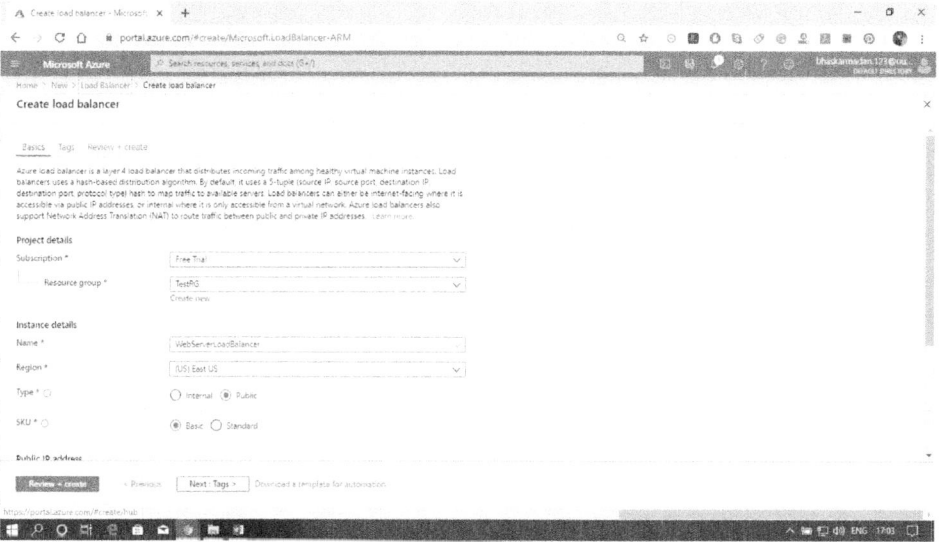

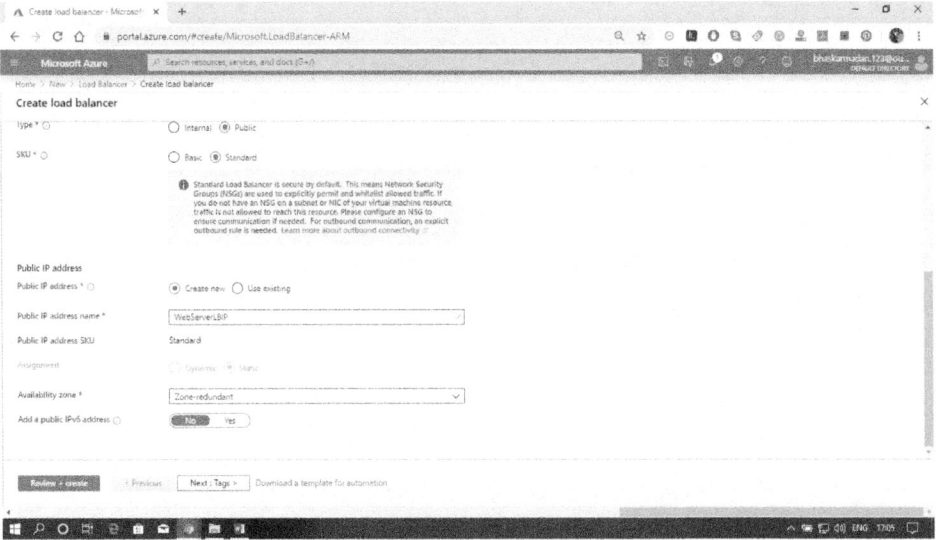

Step 3: You will be redirected to the review page. Check all the details and click on Create.

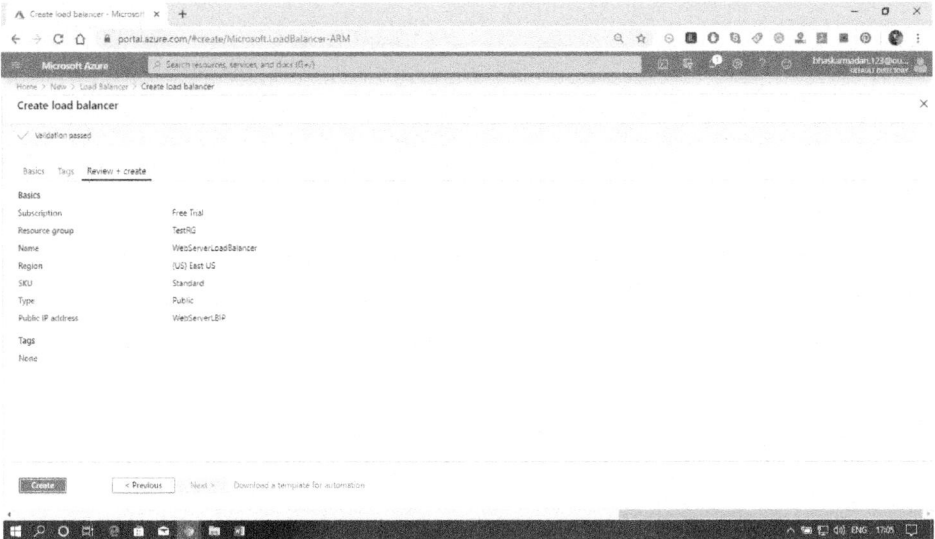

Your Load Balancer is now created.

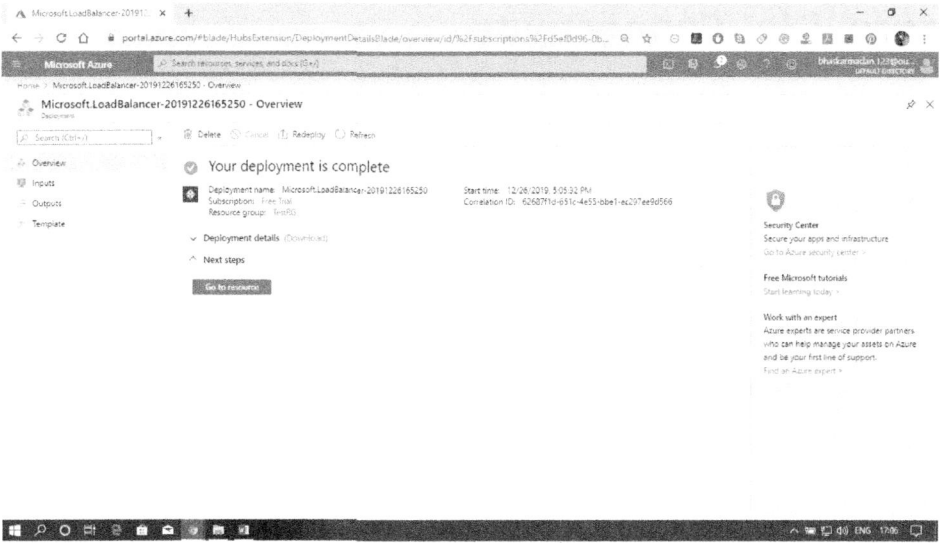

Azure VNet Connectivity

In a typical IT environment, we tend to have multiple virtual networks, and also the workloads in these different virtual networks need to communicate with each other. So, we will discuss some connectivity scenarios which we can use to enable communication between workloads in various virtual networks.

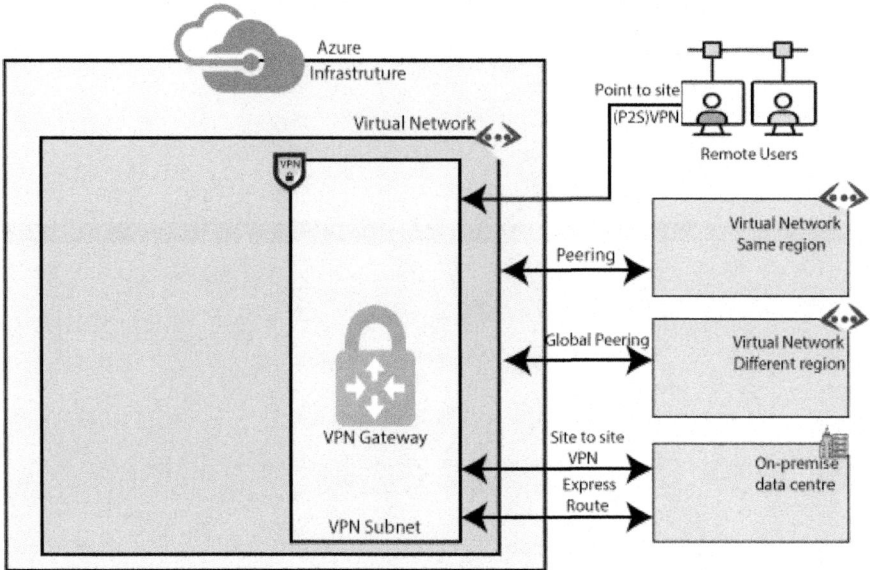

Peering

Virtual network peering enables us to connect two VNet in the same or across regions. If both of the virtual networks are in Azure and also within the same region, then you can use peering. Due to this, the workload in those virtual machines can communicate with each other.

❖ The traffic between different virtual machines in peered virtual networks is routed directly from the Microsoft backbone infrastructure, not through a gateway or over the public Internet.
❖ We can deploy hub-and-spoke networks, where the virtual hub network can host infrastructure components such as a virtual network appliance or a VPN gateway.
❖ Every spoke virtual network could then peer with the hub virtual network Traffic can flow through virtual network appliances or VPN gateways in the virtual hub network.

* When virtual networks have peered, we can also configure the gateway in the peered virtual network as a transit point to an on-premises network.

Global peering

If we have a virtual network in azure that exists in different regions, then we can use Global peering. Gateway transit is supported for both VNet peering and Global VNet Peering.

Site to Site VPN: If we have an on-premises virtual network, and we may have other virtual networks existing in other cloud providers. To connect to our virtual network in Azure with the network that is an on-premises data center, we can use Site to site VPN.

Express Route: If we have a business requirement where this connection between our on-premises data center and virtual network should be on a private channel of communication, then you can use Express Route.

Points to Remember while peering:
* Peering between VNets is allowed in the same subscription only.
* Peering between VNets in different subscriptions under the same AD tenant is allowed.
* Peering between VNets in different subscription located in different AD tenants are also allowed.

VPN Gateway

A VPN gateway is a specific type of virtual network gateway, which is used to send encrypted traffic between an Azure virtual network and an on-premises location over the public internet. VPN gateway act as a middle man on both sides of the virtual networks. And if the workloads in those virtual networks need to communicate with each other, they will communicate via this encrypted channel of communication between the VPN gateways of both virtual networks.

When we are planning to deploy a VPN gateway into Azure, we can configure the number of setting related to it:

* **Gateway SKUs:** We need to select the SKU that satisfies our requirements based on the types of workloads, throughputs, features, and SLAs.
* **Zone-redundant gateways:** We can get benefits from zone-resiliency to access your mission-critical, scalable service on Azure when we use zone-redundant gateways.
* **Connection types:** Connection type can be IPsec, Vnet2Vnet, ExpressRoute, VPNClient.

❖ **VPN types:** The VPN type that we choose depends on the connection topology that we want to create and the VPN device. It can be a policy-based VPN or Route-based VPN.

❖ **Gateway subnet:** Before you create a VPN gateway, you must create a gateway subnet with the name 'GatewaySubnet' and do not deploy anything else into that subnet.

❖ **Local network gateway:** Local network gateway usually represents your on-premises location, i.e., VPN devices, and address prefixes.

❖ **Connection topologies:** Site to site, Multi-site, point-to-site, Vnet-to-Vnet, and express route.

❖ **Monitoring and Alerts:** Monitors the key metrics and configure alerts.

Section IV: Azure Compute Services

Azure Compute Service

The word compute here refers to the hosting model for the computing resources on which our application runs. Azure compute service can be divided broadly into three categories.

- ❖ Infrastructure as a service
- ❖ Platform as a service
- ❖ Serveless services

The most fundamental building block is the Azure virtual machine. Using Azure virtual machine, we can able to deploy different services such as Windows, Linux within the Azure cloud. When we implement a virtual machine, every virtual machine will have an associated OS and data disk.

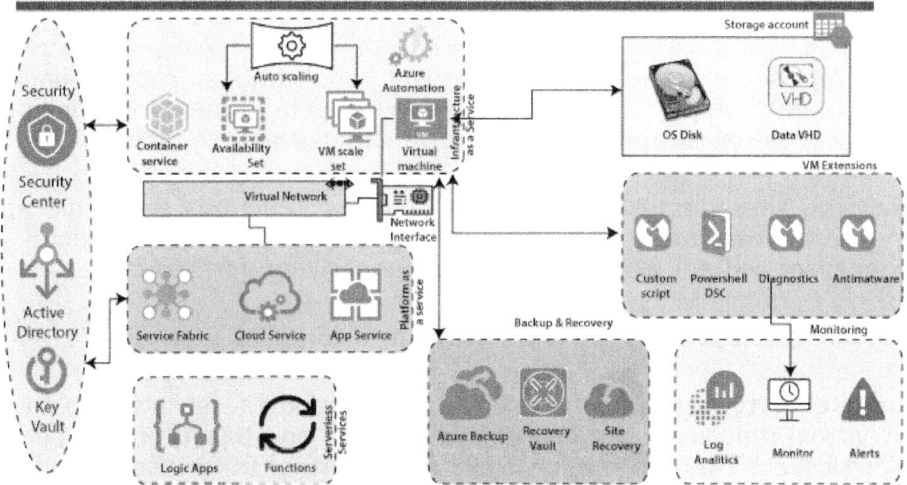

Azure compute building blocks

Azure compute options
Following are the main compute options available in Azure:

- ❖ **Virtual Machine:** It is an IaaS service, allowing us to deploy and manage VMs inside a virtual network (VNet).
- ❖ **App Service:** It is a managed PaaS offering for hosting web apps, mobile app back ends, RESTful APIs, or automated business processes.
- ❖ **Service Fabric:** It is a platform that can run on any environment, including Azure or on-premises. It is an orchestrator of micro-services across a cluster of machine
- ❖ **Azure Kubernetes Services:** It manages a hosted Kubernetes service for running containerized applications.
- ❖ **Azure Container Instances:** It offers the fastest and most straightforward way to run a container in Azure without having to provision any virtual machines and without having to adopt a high-level service.
- ❖ **Azure Functions:** It is a managed FaaS service.
- ❖ **Azure Batch:** It is a managed service for running large-scale parallel and high-performance computing (HPC) applications.
- ❖ **Cloud Services:** It is a managed service for running cloud applications. It uses a PaaS hosting model.

When you are deploying any virtual machine, such as running some scripts, etc.. For that purpose, Azure provided several extensions such as custom script, PowerShell DSC, which stands for desired state configuration. You can have a diagnostic extension to collect all the logs that are emitted from that virtual machine. Also, we can have anti-malware software installed on that virtual machine to protect against viruses, etc.

Using App service, we can deploy web applications, mobile backend services, API Apps, etc. If we have a requirement to deploy a microservices-based application, then we can use service fabric.

Within the serverless service, we have Azure functions and logic apps. Using which we can able to deploy snippets of code on the cloud and trigger them without worrying about the underlying infrastructure.

Three key services are associated with Azure compute service:

Azure security center: It is used to understand the security posture of your virtual machines. We can define policies, and based on the policies, we can collect the information from Azure virtual machines and identify the threat. It will provide recommendations associated with that.

Active Directory: It is used to control that who can access virtual machines or scale sets or availability sets or in-fact any other Azure services within Azure.

Key Vault: It is used to store certificated keys or any sensitive information within Azure securely.

Azure Virtual Machines

Azure Virtual machine will let us create and use virtual machines in the cloud as Infrastructure as a Service. We can use an image provided by Azure, or partner, or we can use our own to create the virtual machine.

Virtual machines can be created and managed using:

- ❖ Azure Portal
- ❖ Azure PowerShell and ARM templates
- ❖ Azure CLI
- ❖ Client SDK's
- ❖ REST APIs

Following are the configuration choices that Azure offers while creating a Virtual Machine.

- ❖ Operating system (Windows and Linux)
- ❖ VM size, which determines factors such as processing power, how many disks we attach etc.
- ❖ The region where VM will be hosted
- ❖ VM extension, which gives additional capabilities such as running anti-virus etc.
- ❖ Compute, Networking, and Storage elements will be created during the provisioning of the virtual machine.

VM Sizes

It is important to select the right VM size and type for the working of our virtual machine perfectly. So, these are the VM sizes that are available within Azure.

Type	Sizes	Description
General-purpose	B, Dsv3, Dv3, DSv2, Dv2, DS, D, Av2, A0-7	It has balanced CPU-to - memory ratio, It is ideal for testing and development, small to medium databases, and low to medium traffic web

		servers.
Compute-optimized	Fsv2, Fs, F	It has a high CPU-to-memory ratio. It is suitable for medium traffic web servers, network appliances, batch processes.
Memory-optimized	Esv3, Ev3, M, GS, G, DSv2, DS, Dv2, D	Is has a high memory-to-CPU ratio. Great for relational database servers, medium to large caches, and in-memory analytics.
Storage optimized	Ls	It has high disk throughput and IO that is Ideal for Big Data, SQL, and NoSQL databases.
GPU	NV, NC, NCv2, ND	It is a specialized virtual machine that is targeted for heavy graphic rendering and video editing. Available with single or multiple GPUs.
High performance	H, A8-11	It is the fastest and most powerful CPU virtual machine with optional high-

compute	throughput network interfaces (RDMA).

Creating Azure Virtual machine in Azure Portal

Step 1: Click on All services and then click on the Virtual machine button, as shown in the following image.

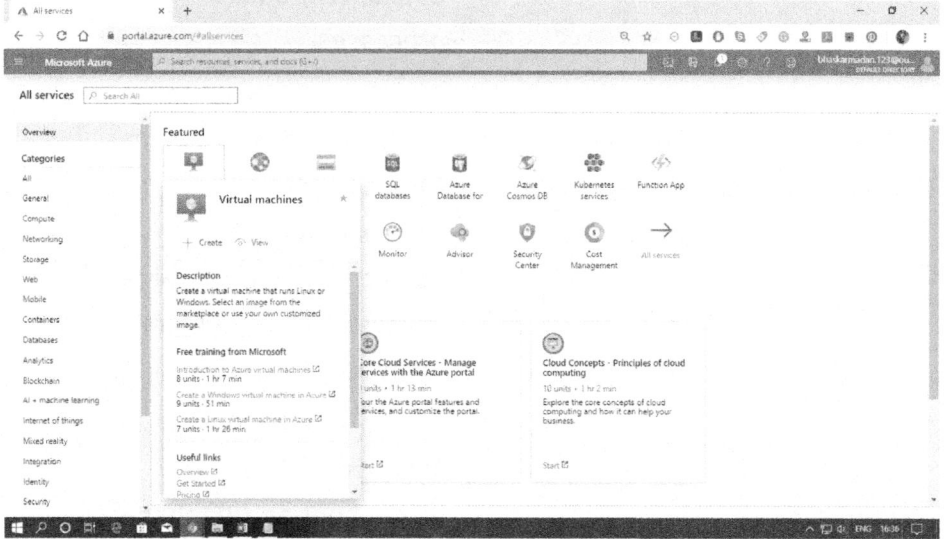

Step 2: Click on create, then you will be redirected to the Create Virtual machine page.

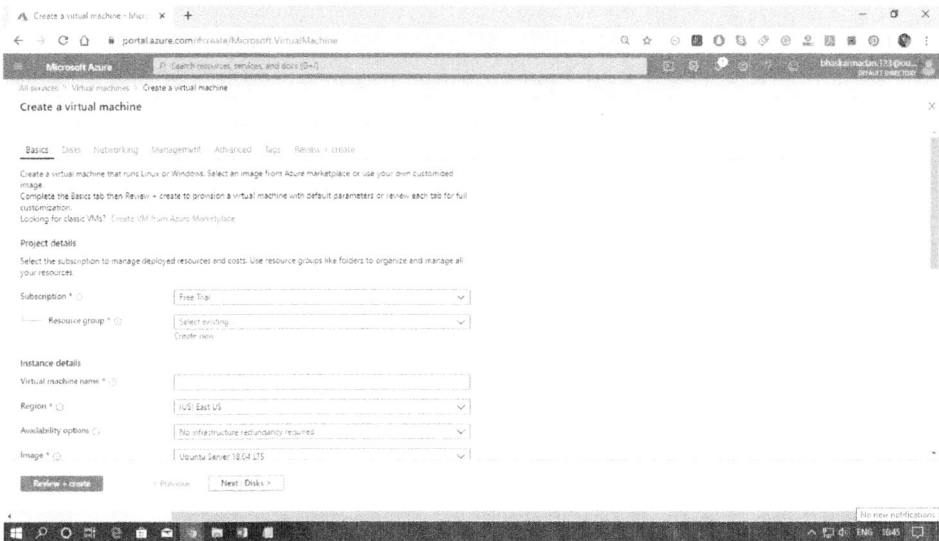

Step 3: Now, select the image for your virtual machine from the Azure marketplace by clicking on "Create VM from Azure marketplace".

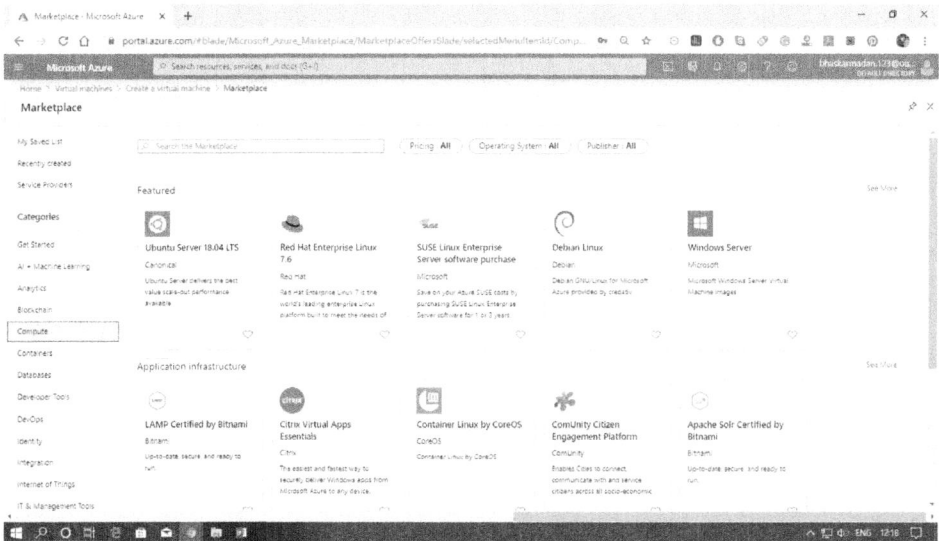

Step 4: After selecting the image, the first thing you need to do is to provide a name to your virtual machine.

[113]

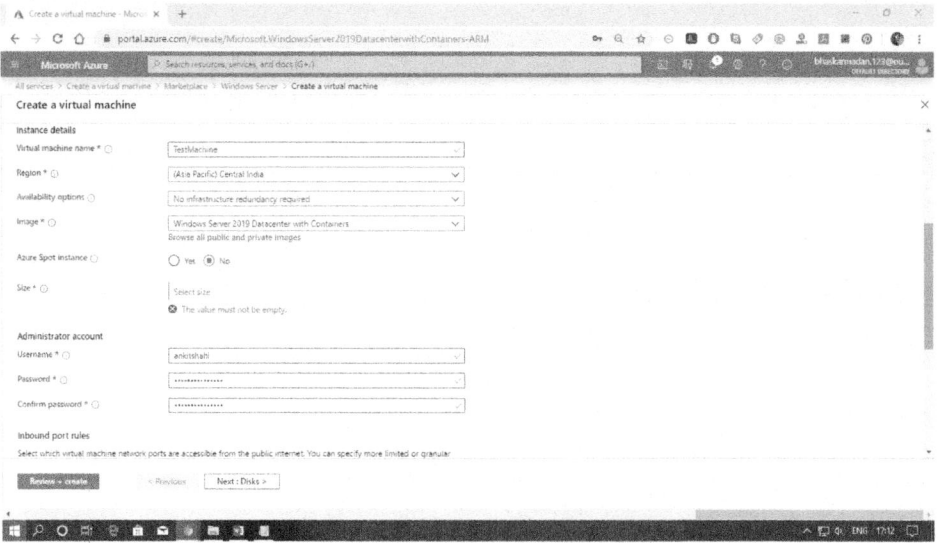

Step 5: Now select the size and type of VM according to your requirements. After that, set a User name and password for your Virtual Machine then click next.

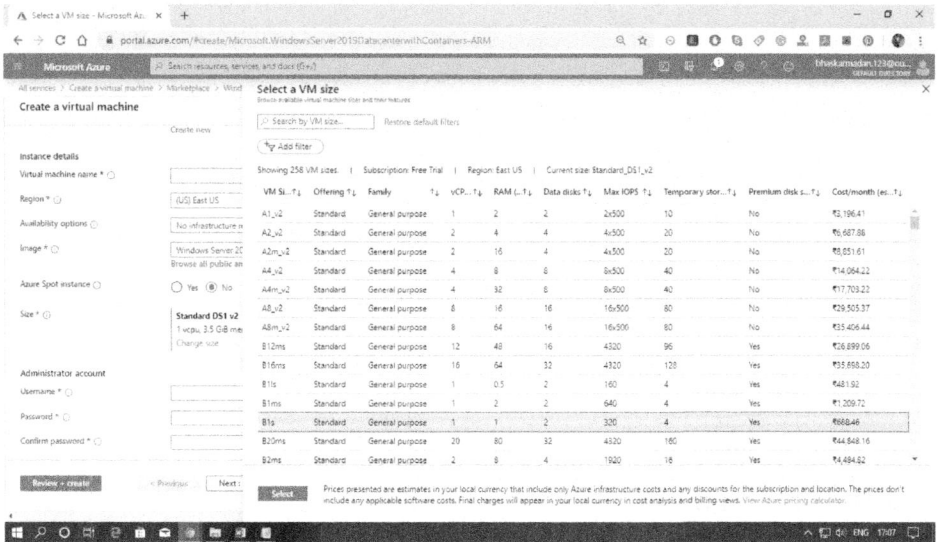

Step 6: You are on the disk tab now, Select the disk type you need then click next to redirect on the networking page.

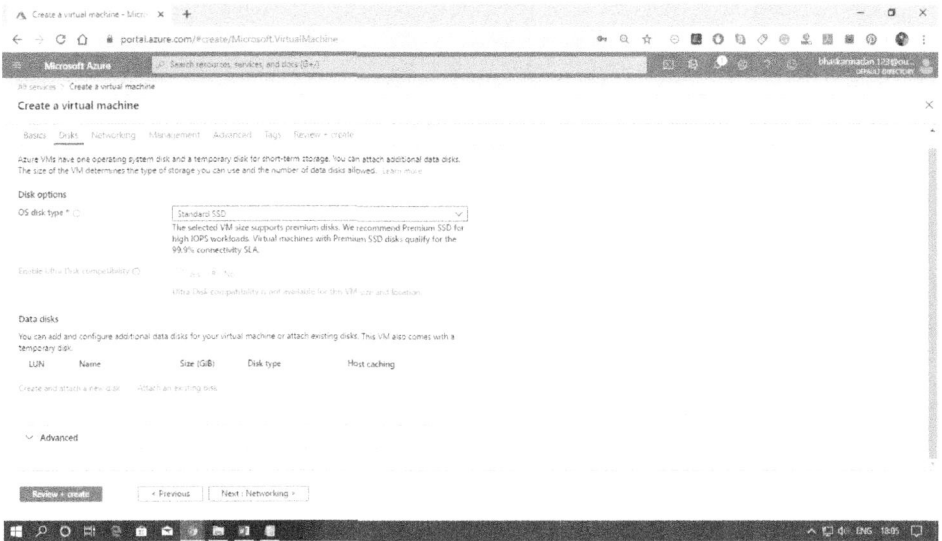

Step 7: Select the virtual network, subnet, and IP address for the Virtual machine. We are leaving it as default because we are creating it for the training purpose.

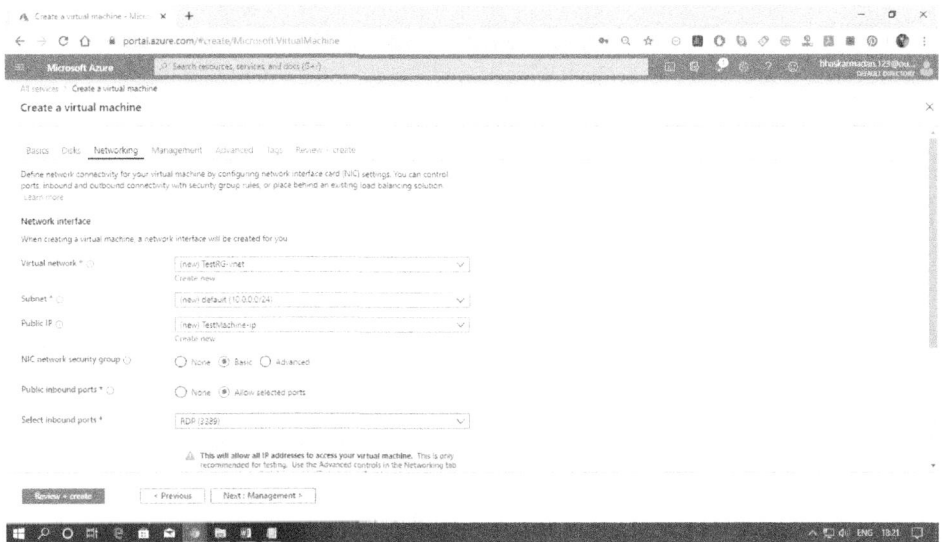

Step 8: Now select the management tab, and choose the boot and OS diagnostic option. Then click next.

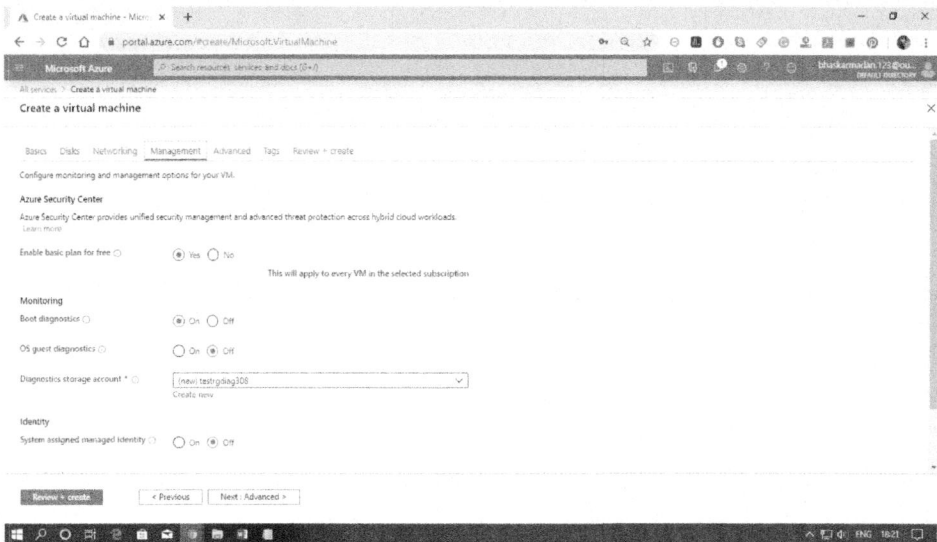

Step 9: In advanced settings, you can embed an extension to the virtual machine.

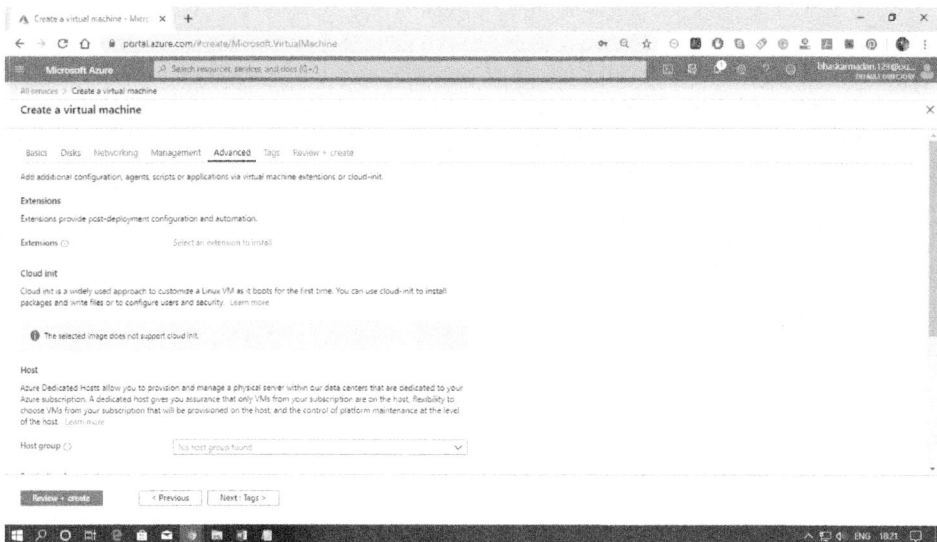

Step 10: Now, on the review and create a window, click on the create button.

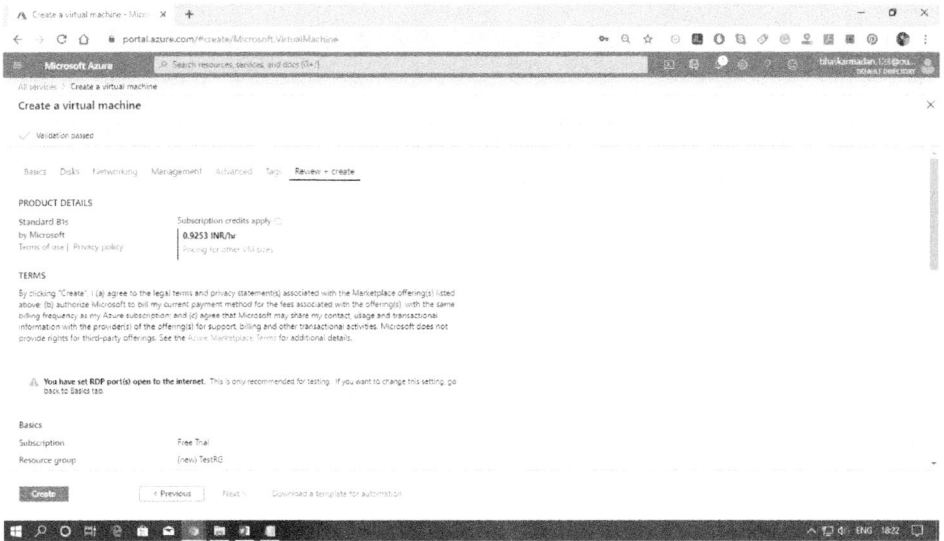

Step 11: When you click on Create, the further process will start. Wait for a few minutes to complete the processing.

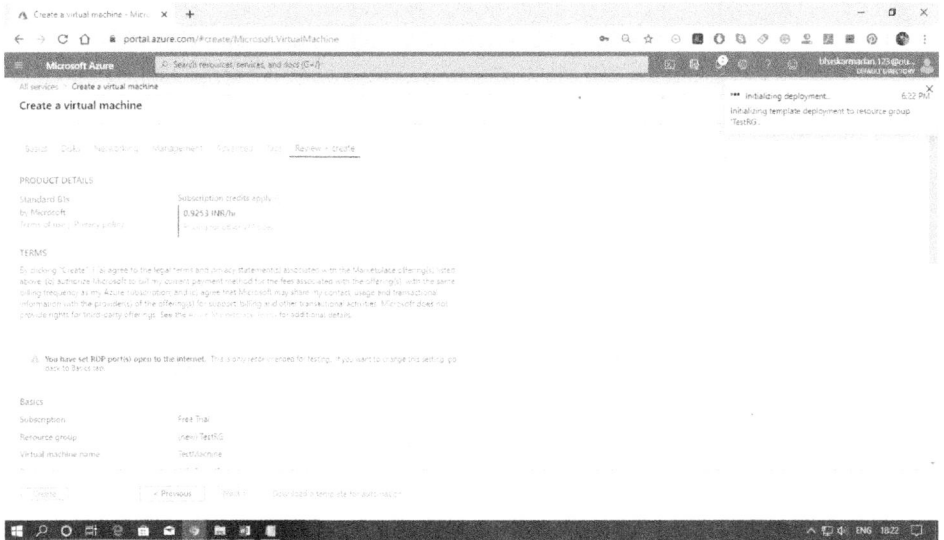

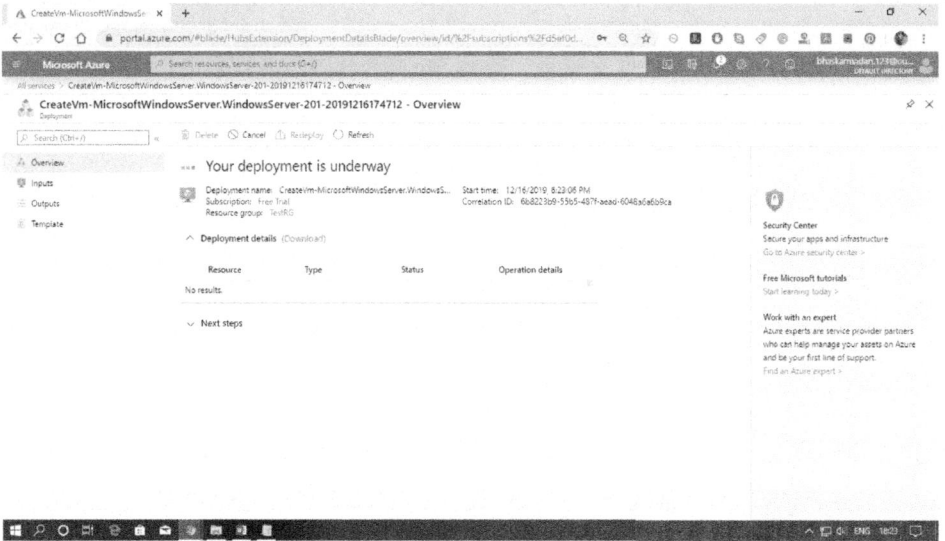

When your Virtual Machine is created, the following window will appear. You can now use your virtual machine.

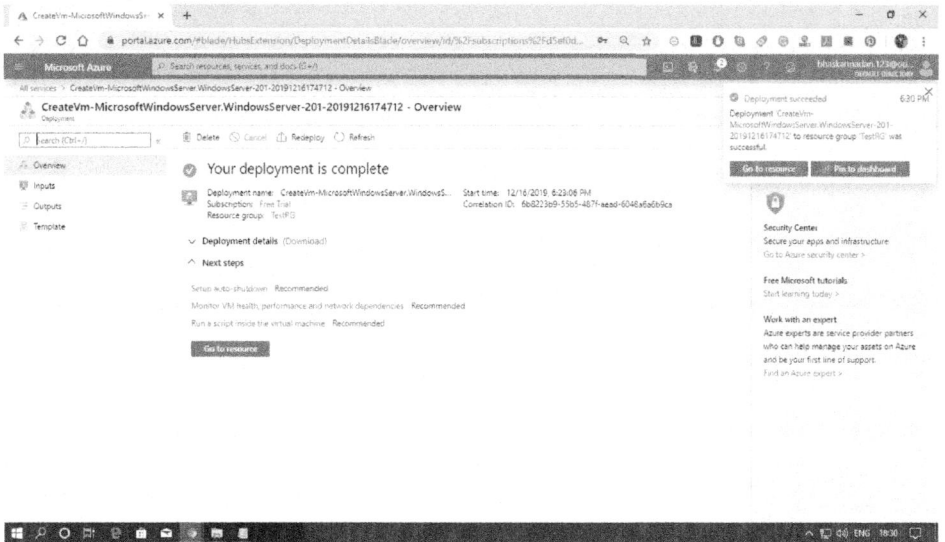

Azure VM Storage

Disks used by VMs

❖ **Operating system disk:** In Azure, every virtual machine will have an operating system disk.

❖ **Temporary disk:** Each VM contains a temporary drive. The temporary disk provides short-term storage for applications and processes.

❖ **Data disk:** A data disk is a VHD that's attached to a virtual machine to store application data or other data we need to keep.

Performance tiers

❖ **Standard Storage:** It is backed by HDDs and deliver cost-effective storage while still being performant. It is ideal for development and testing, not-critical, and Infrequent access because the max throughput and IOPS per disk is 60MB/s and 500, respectively.

❖ **Premium Storage:** It is backed by SSDs, and deliver high-performance, low-latency disk support for VMs running I/O-intensive workloads. The maximum throughput and IOPS per disk are 250MB/s and 7500, respectively.

Types of the disk in Azure

❖ **Unmanaged disks:** It is the traditional type of disks that have been used by VMs. We can create our storage account and specify the storage account when you create the disk. The scalability targets of SA (20, 000 IOPS) are not exceeded.

❖ **Managed disks:** Managed disks handles the storage account creation/management. We do not have to care about scalability limits the storage account. Microsoft always recommends us to use Azure Managed Disks for new VMs.

Disk encryption

❖ **Storage Service Encryption:** Azure Storage Service Encryption provides encryption-at-rest and safeguards our data to meet our organizational security and compliance commitments. It is enabled by default for all Managed Disks, Snapshots, and Images in every region where managed disks are available.

❖ **Azure Disk Encryption:** Azure Disk Encryption allows you to encrypt the OS and Data disks used by an IaaS Virtual Machine. For Windows, the drives are encrypted using industry-standard BitLocker encryption technology. For Linux, the disks are encrypted using the DM-Crypt technology.

Virtual Machine Availability

❖ **Availability Set:** Availability Set is a logical grouping of VMs within a data center that allows Azure services to understand how our application is built to provide redundancy and availability. An availability set is composed of two additional groupings that protect against hardware failures and allow updates to be applied safely.

　　❖ **Fault domains-** It is a logical group of the underlying hardware that shares a common network switch and power source, similar to a rack within an on-premises datacenter.

　　❖ **Update Domain:** It is a logical group of the underlying hardware that will go under maintenance or be rebooted at the same time.

　　❖ **Managed Disk fault domains:** For VMs using Azure Managed Disks, VMs are aligned with managed disk fault domains when using a managed availability set. This alignment ensures that all the managed disks attached to a VM are within the same managed disk fault domain.

❖ Availability Zones: It is a physically separate zone within an Azure region. There are three Availability zones per supported within the Azure region. All availability zone has the same amount of power source, network, and cooling, and is separated from the other Availability Zones within the Azure region.

Storage Availability
Azure Managed Disks

❖ **Locally redundant storage (LRS):** We will have three copies of the same data within the same facility. So, if there is a datacenter failure, then there is a high probability that we might lose the data.

Storage account-based disks

❖ Locally redundant storage (LRS): It maintains three replicas with the facility.

❖ Zone redundant storage (ZRS): It maintains three replicas but across facilities.

❖ Geo-redundant storage (GRS): The replicas will be maintained in a paired region. For example - if our disk is in Central US, a copy will be kept in East US also.

❖ Read-access geo-redundant storage (RA-GRS): The copy will be available for read-only access in a different region.

Creating Availability Set

We will see here how Azure evenly distributes your virtual machines into different fault and update domains of that availability set.

Step 1: Click on New, then type in an Availability set, and press enter. Now, Click on Create.

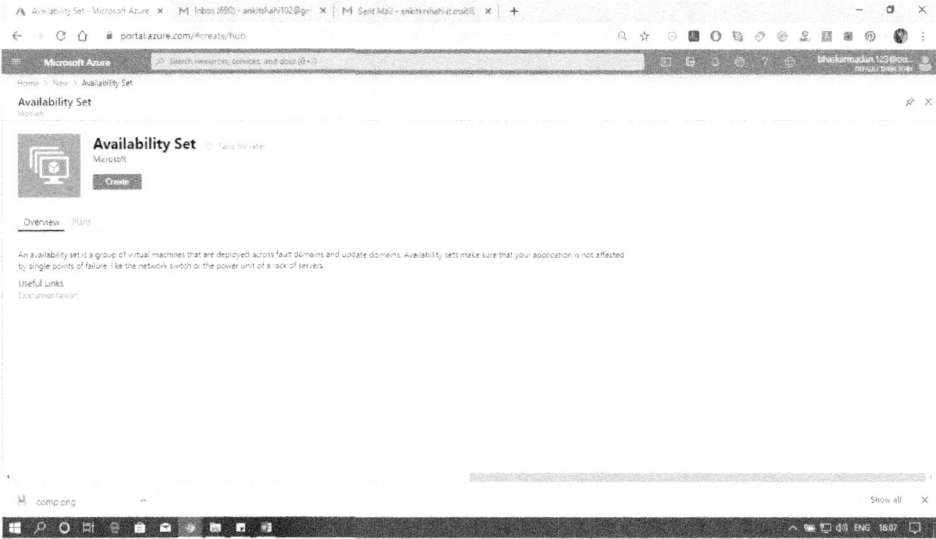

Step 2: You are currently on the Availability set creation page. Fill in all the required details as shown in the figure below, and click on Create.

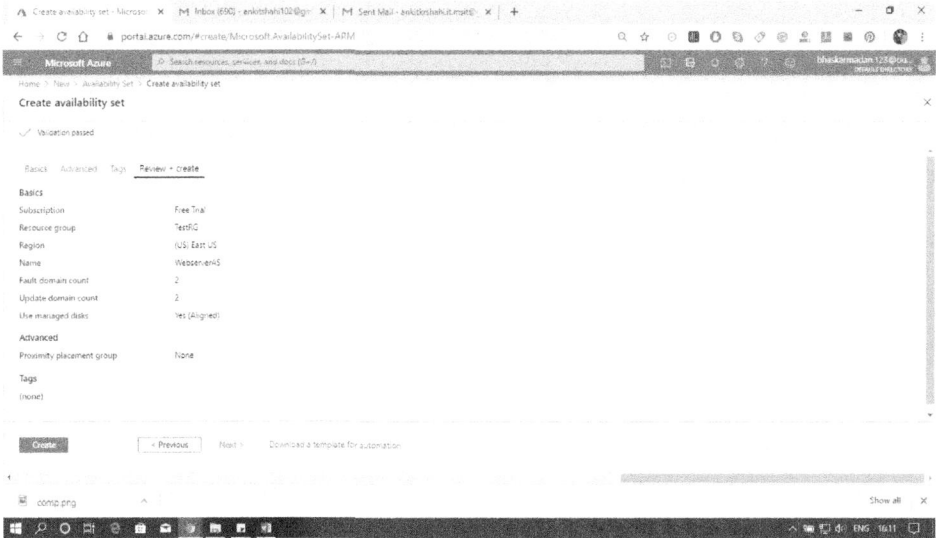

Step 3: Now Click on Go to resource to open the Availability set.

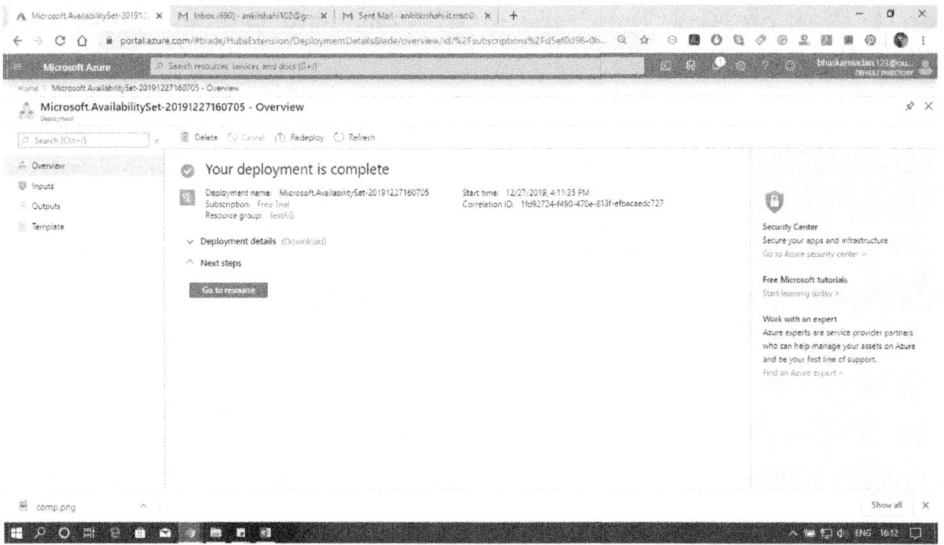

Step 4: Click on Virtual Machine. Any virtual machine that you have added to this Availability set will show here.

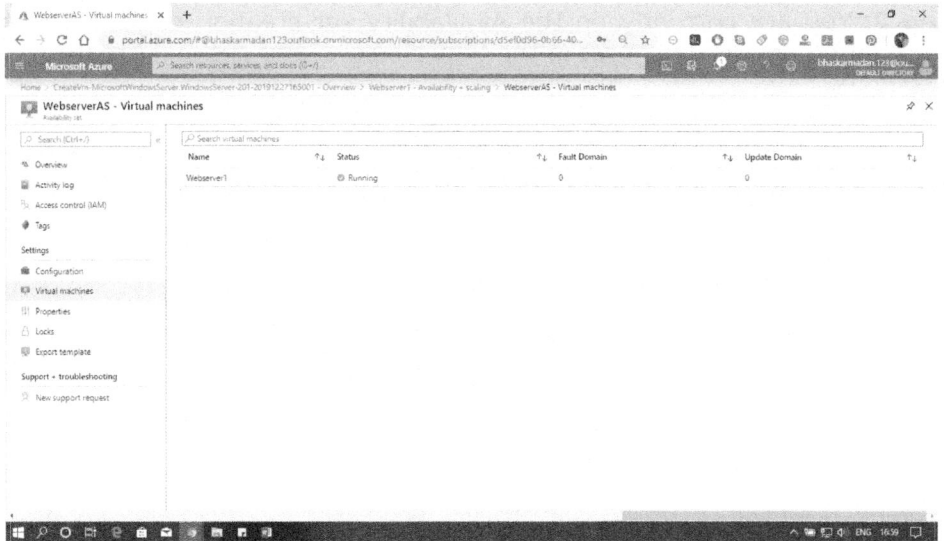

This is how you can make a web server farm tolerant of both unplanned outages and planned outages.

Azure Virtual Machine Scale Set & Auto Scaling

Virtual Machine scale sets

The scale sets are Azure compute resources that can be used to deploy and manage identical VMs. They are designed to support virtual machine auto-scaling. VM scale sets can be created using the Azure portal, JSON templates, and REST APIs. To increase or decrease the number of VMs in the scale set, we can change the capacity property and redeploy the template. A virtual machine scale set is created inside VNET, and individual VMs in the scale set are not allocated with public IP addresses.

Any virtual machine that we deploy and is the part of the virtual machine scale set will not be allocated with a public IP address. Because sometimes, the virtual machine scale set will have a front end balancer that will manage the load, and that will have a public IP address. We can use that public IP address and connect to underlying virtual machines in the virtual machine scale set.

Virtual Machine Auto Scaling

Autoscale enables us to dynamically allocate or remove resources based on the load on the services. You can specify the maximum and the minimum number of instances to run and add or remove VMs based on a set of rules within the range.

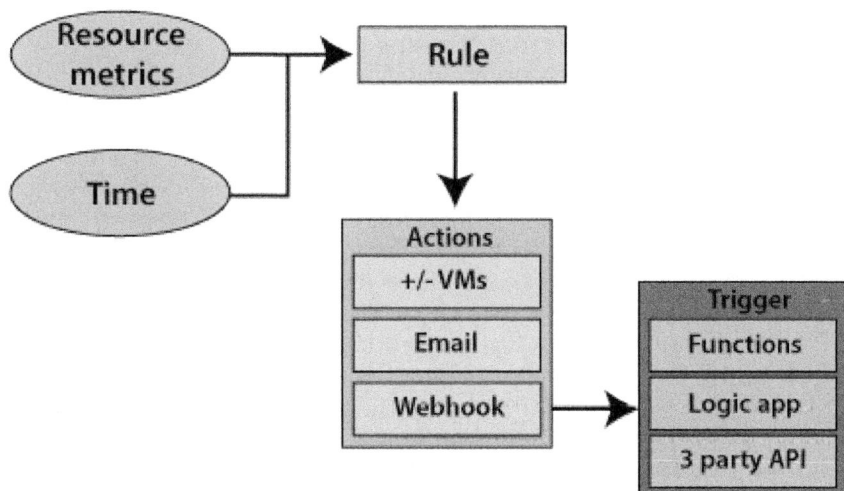

Resource metrics → Rule

Time

Rule → Actions
Actions: +/- VMs, Email, Webhook

Trigger: Functions, Logic app, 3 party API

The first step in auto-scaling is to select a metric or time. So, it can be a metric based auto-scaling, or it can be a schedule based auto-scaling. The metrics can be CPU utilization, etc., and the time can be like the night at 6 o'clock till morning 6:00, we want to reduce no of servers. We can have a schedule based auto-scaling. In case if we're going to reach according to load, then we can use metric based auto-scaling.

The next step in the auto-scaling is to define a rule with the condition. For example - if the CPU utilization is higher than 80 percent, then spin off a new instance. And once the condition is met, we can carry some actions. The actions can be adding or removing virtual machines, or it can be sending email to a system administrator, etc. We need to select whether it is a time-based auto-scaling or metric-based, and we need to choose the metric. We define the rule and actions that need to be triggered when the condition in that rule is satisfied.

Horizontal and Vertical scaling

❖ **Horizontal scaling:** The increasing or decreasing the number of VM instances. It auto-scales horizontally and sometimes called as Scale-out or Scale in scaling.
❖ **Vertical scaling:** In this, we keep the same numbers of VMs but make VM more or less powerful. Power is measured as memory, CPU speed, disk space, etc. It is limited by the availability of larger hardware within the same region and usually requires a VM to start and stop. This is sometimes called Scale up or scale downscaling. Below are the steps to achieve vertical scaling.
o Setup Azure automation account
o Import the Azure Automation Vertical scale runbooks into our subscriptions.
o Add a webhook to our network.
o Add an alert to our Virtual Machine.
o We can also scale web apps and cloud services.

Metrics for Autoscaling

❖ **Compute metrics:** The available metrics will depend upon the installed operating system. For windows, we can have a processor, memory, and logical disk metrics. For Linux, we can have processor, memory, physical & network interface metrics.
❖ **Web Apps metrics:** It includes CPU & memory percentage, Disk & HTTP queue length, and bytes received/sent.
❖ **Storage/ Service bus metrics:** We can scale by Storage queue length, which is the number of messages in the storage queue. Storage queue length is a particular metric, and the threshold applied will be the number of messages per instance.

Tools to implement Auto Scale

❖ We can use the **Azure portal** to create a scale set and enable auto-scaling based on a metric.

- ❖ We can provision and deploy VM scale sets using **Resource Manager Templates**.
- ❖ **ARM templates** can be deployed using Azure CLI, PowerShell, REST, and also directly from Visual Studio.

Scaling Azure Virtual Machine

Step 1: Go to Azure Marketplace and type in the Virtual Machine scale set. Then click on Create.

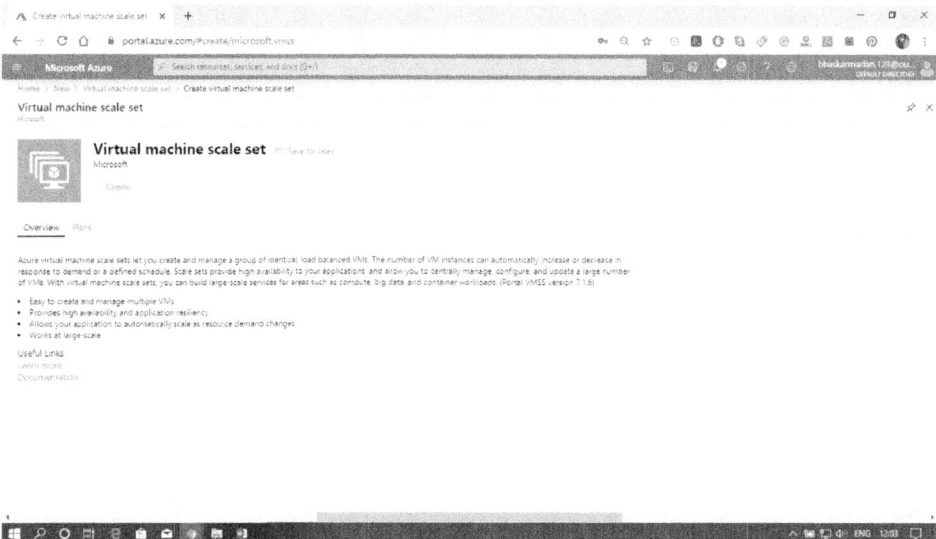

Step 2: We need to give a name to this scale set. And fill all the other required details, as shown in the figure below. Then click on create.

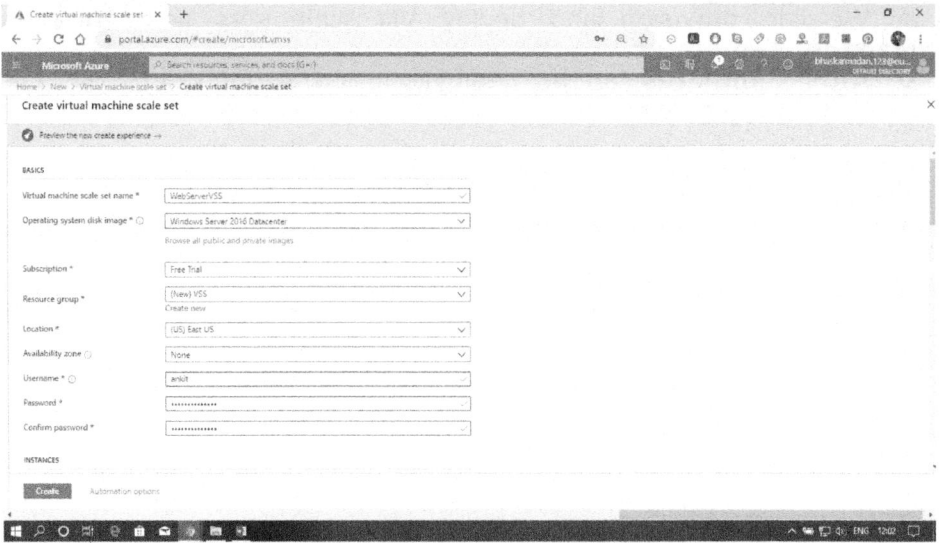

Step 3: Now, your Virtual Machine scale set is successfully deployed. To view VMSS, you can go to resources.

Step 4: Now, click on scaling. Provide an auto-scale setting name. And select the resource group.

Step 5: Scroll Down, and you will find two ways to auto-scale. First, click on "add a rule? for auto-scaling based on the metric. We are going to scale our virtual machine if the average percentage of CPU utilization is above 70 percent.

Step 6: Now, select the time and date based scaling, where you can scale when you need more space. And the last thing is Notify, where you get notified whenever the auto-scaling gets triggered.

Azure Backup

Azure Backup is a service provided by Microsoft Azure to back up and restore our data over the Microsoft cloud. Azure Backup replaces our existing on-premises or off-site backup solution with a cloud-based solution that is reliable, secure, and cost-competitive. It is not only used as cloud storage to back-up our data to the cloud, but we can also use our existing local disc to back-up the data.

So, Azure backup works with this heterogeneous storage environment with the combination of on-premises storage and also cloud storage. Whenever Azure backup uses local storage, we'll not get charged for it. You will only get charged when the data is backed up in the cloud.

Advantages of Azure Backup
❖ Automatic storage management
❖ Unlimited scaling
❖ Multiple Storage options
❖ Application consistent backup
❖ Long-Term retention

Working of Azure backup service

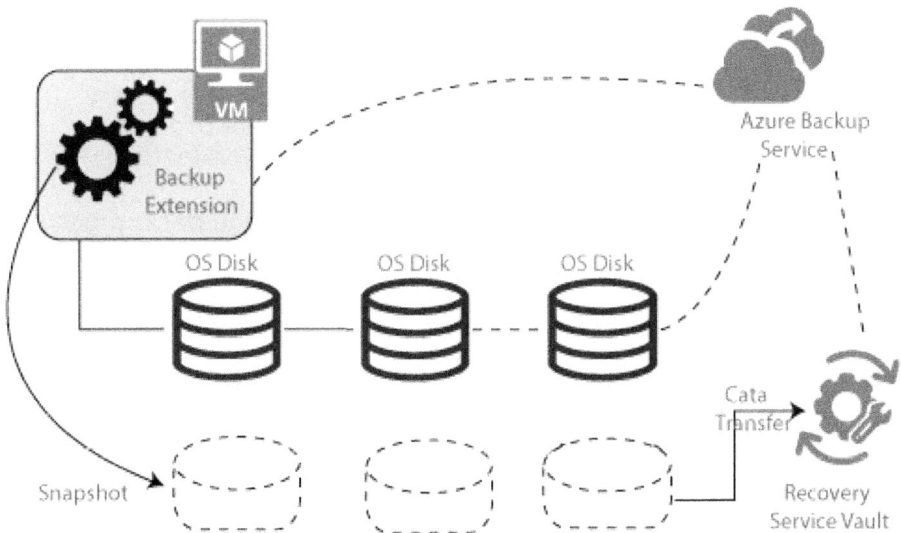

The first thing we do when we're using Azure backup is to define backup policy. The policy describes how frequently you need to take a backup and also which target you need to backup. After that, we also identify the destination where the data need to get stored, which in most cases, it's going to be recovery services vault. Backup of the data will not get stored in a storage account, and it will be stored in the recovery service vault, which is also an online storage facility where we can save the backup of all our virtual machines.

Components of Azure Backup service

Component	What is protected?	Backup storage	Backup frequency
Azure Backup agent	Files, Folders, System State Windows only.	Recovery service vault	Three backups per day
System Center DPM	Files, Folders, Volumes, VMs, Applications, Workloads	Recovery service vault, Local disk, Tape	Two backups per day to RSV, Every 15 minutes for SQL Server, Every hour for other workloads
Azure Backup Server	Files, Folders, Volumes, VMs, Applications, Workloads	Recovery service vault, Local disk	Two backups per day to RSV, Every 15 minutes for SQL Server, Every hour for other workloads
Azure IaaS VM Backup	VMs, All disks Windows and Linux	Recovery service vault	One backup per day

How to take the backup of VM using the Azure backup

Step 1: Go into Virtual Machine and click on Backup. The following window will appear once you click on backup.

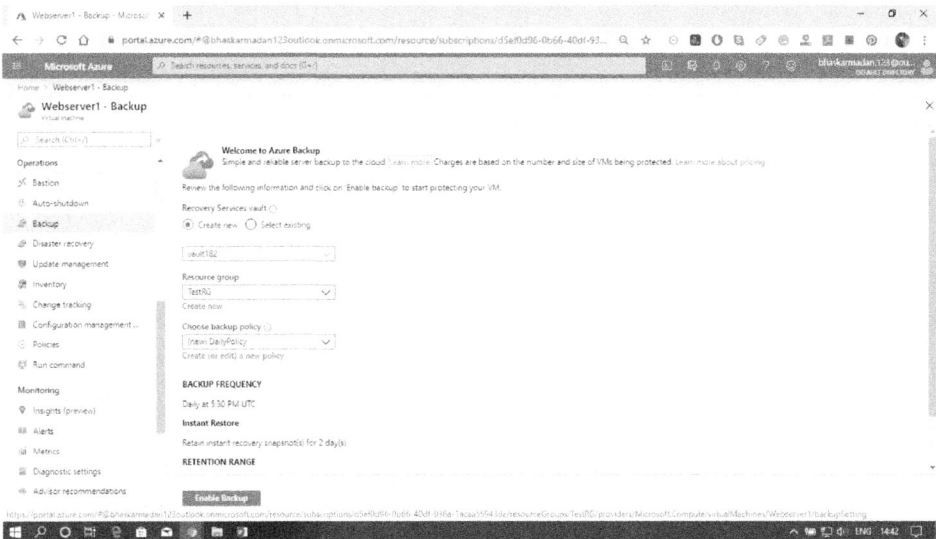

Step 2: Configure the Backup policy according to your requirements and Click on Enable Back up.

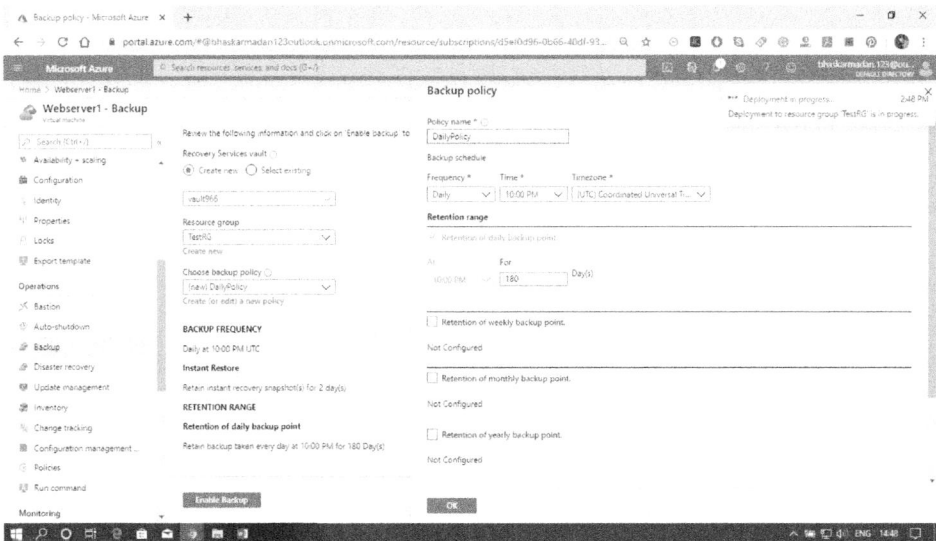

Step 3: Now, go to the resource group and click on the Recovery Service vault that you have created, as shown below.

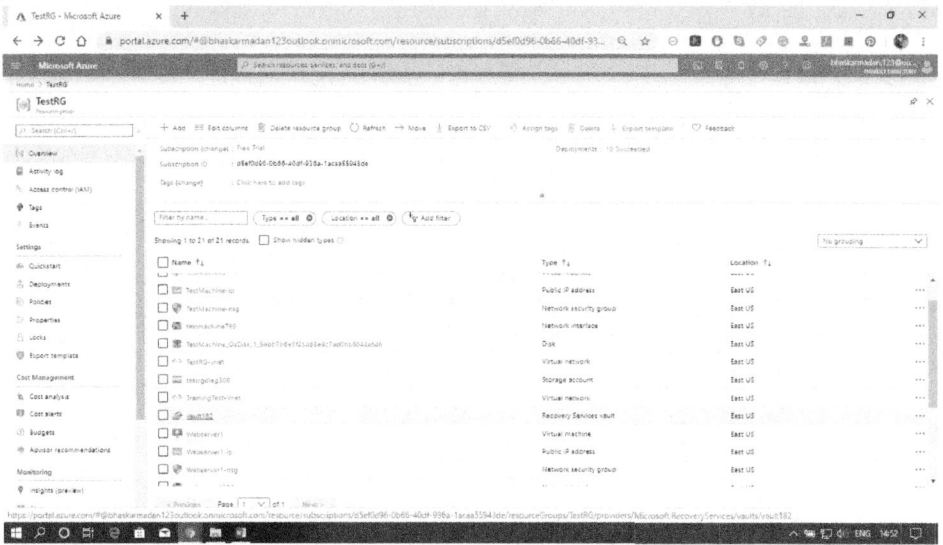

Step 4: In case you would like to take a backup from the recovery services vault and choose different settings, then you can click on backup and then select where your workload is running.

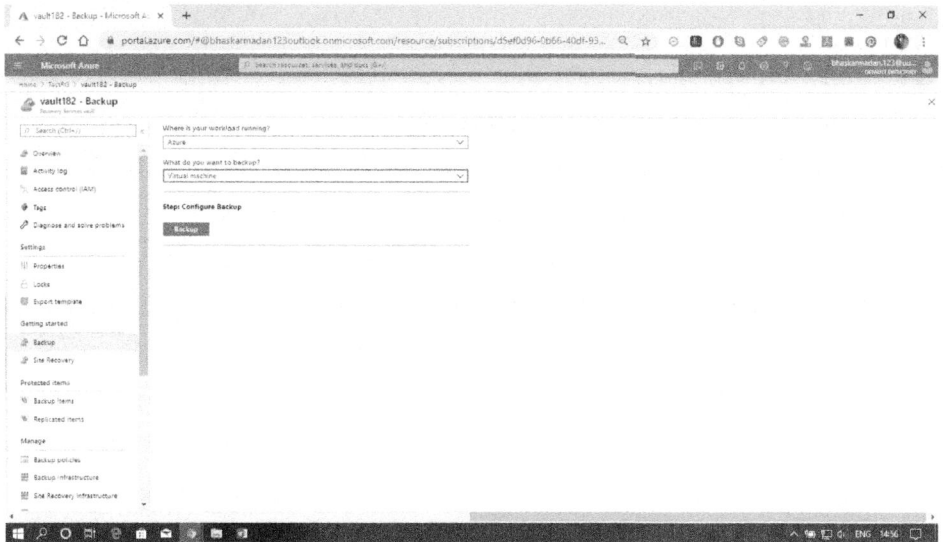

Azure Virtual Machine Security

There are many services available to secure our virtual machine.

Azure Active Directory

❖ By using the Azure Active Directory, we can control access to our virtual machines to different users or groups of users. When we create a virtual machine, we can assign a user to it, and while we are assigning the user to the virtual machine, we will also associate a particular rule to them. That role defines the level of access that the user will have on our virtual machine.

❖ Users, groups, and applications from that directory can manage resources in the Azure subscription.

❖ It grants access by assigning the appropriate RBAC role to users, groups, and applications at a certain scope. The scope of a role assignment can be a subscription, a resource group, or a single resource.

❖ Azure RBAC has three essential roles that apply to all resource types:

❖ **Owner:** They have full access to all resources, including the right to delegate access to others.

❖ **Contributor:** They can create and manage all types of Azure resources but can't grant access to others.

❖ **Reader:** They can only view existing Azure resources.

Azure security center

The Azure security center identifies potential virtual machine (VM) configuration issues and targeted security threats. These might include VMs that are missing network security groups, unencrypted disks, and brute-force Remote Desktop Protocol (RDP) attacks.

We can customize the recommendations we would like to see from the Security Center using security policies.

❖ Set up data collection
❖ Set up security policies
❖ View VM configuration health
❖ Remediate configuration issues
❖ View detected threats

Managed Service Identity

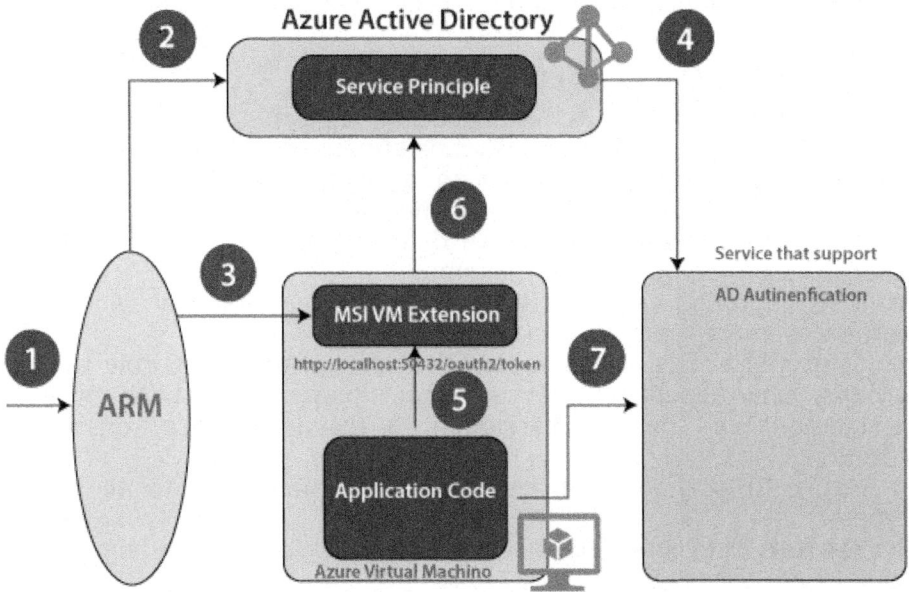

It is newly introduced in Azure. Earlier, what used to happen was whenever we're deploying an application into a virtual machine, we generally have user id and password within a configuration file of a folder of that application. But if someone gets access to that virtual machine, they can be able to go to the configuration file and view that also. To further increase the security of our application code and safety of services that are being accessed by application code, we can use Managed Service Identity.

Other Security Features

❖ **Network security group:** To filter the traffic in and out of the virtual machine.

❖ **Microsoft Antimalware for Azure:** We can install on our Azure virtual machines to secure our machines against any malware.

❖ **Encryption:** We can enable Azure Disk Encryption.

❖ **Key Vault and SSH Keys:** we can use key vault to store the certificates or any sensitive key.

❖ **Policies:** All the security-related policies we can apply using it.

Azure VM Monitoring

There are different Azure services that are available to monitor our Azure virtual machines.

Diagnostics and metrics

❖ Using the activity log, we can monitor and audit the operation carried on a Virtual machine. For example - starting the virtual machine, stopping the virtual machine, reimaging, etc.

❖ Observe base metrics for the VM using Azure monitor. You will see those metrics in the form of a dashboard within the resource section itself. But if you go to Azure monitor, you can monitor all the base metrics of any resources within Azure, including virtual machines.

❖ Enable the collection of boot diagnostics and view it using the Azure portal.

❖ Enable the collection of guest OS diagnostics data, and analyze using OMS (Operation Management Speed).

❖ We can set up and monitor the collection of diagnostics data using metrics in the Azure portal, the Azure CLI, Azure PowerShell, and REST APIs

Alerts

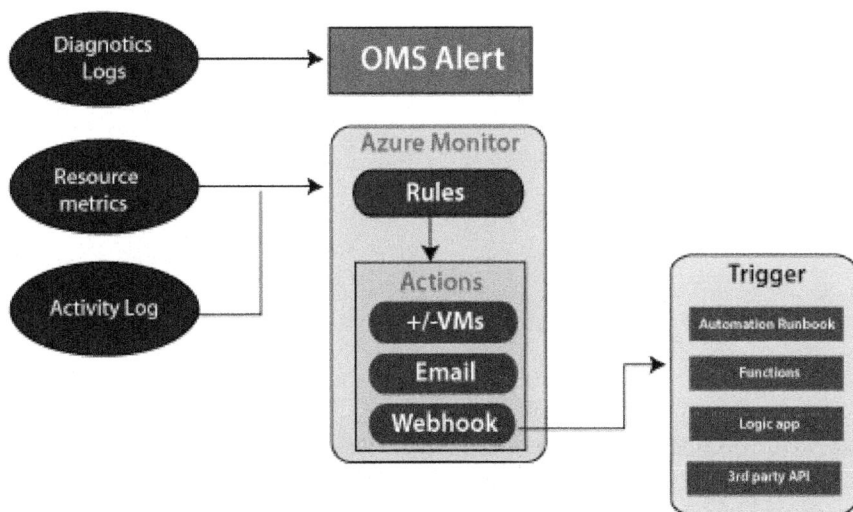

Azure provides a comprehensive ability to get alerted. There are three sources of information against which we can get alerted.

❖ Activity Log

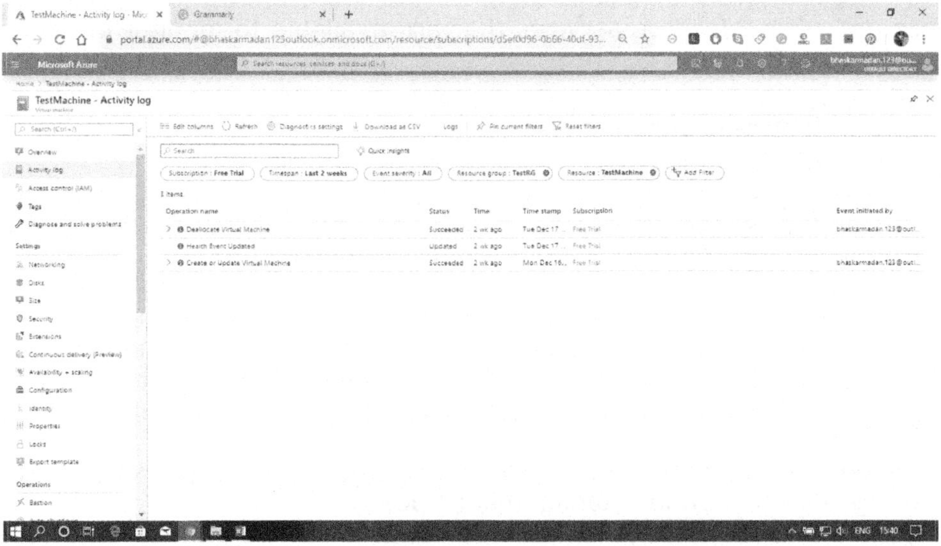

❖ Resource Metrics

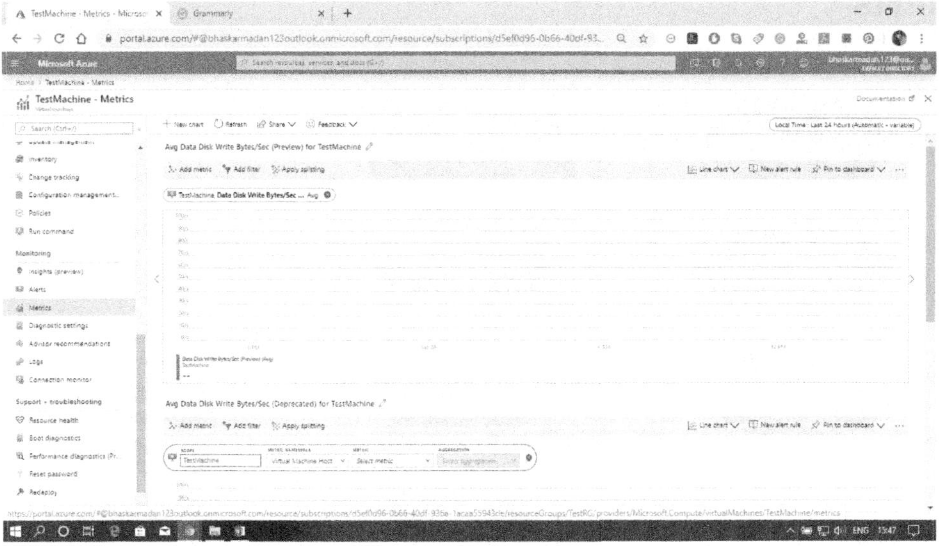

❖ Diagnostic Logs

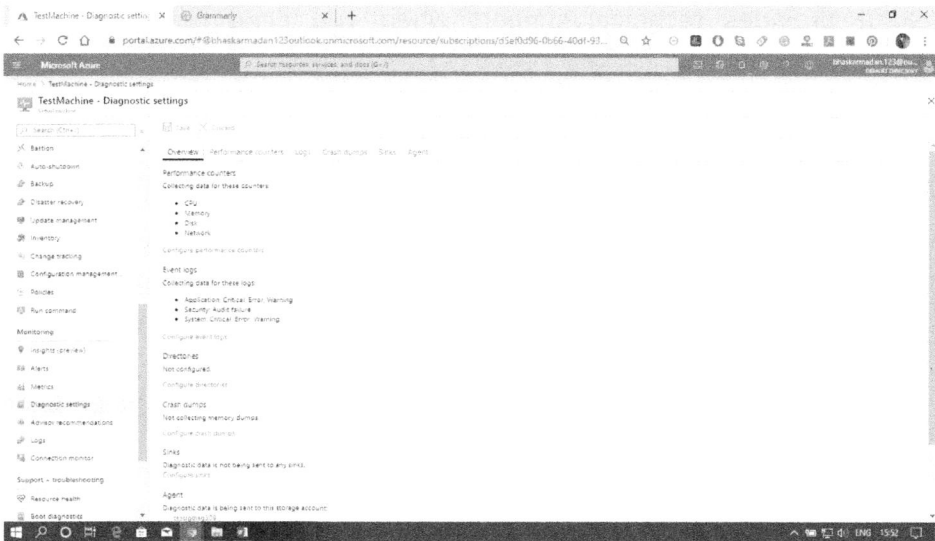

When it comes to diagnostic logs, we can raise an alert based on diagnostic logs using an OMS alert. And in case of resource metrics and activity logs, we'll use the Azure monitor. The azure monitor is a basic one, but it is comprehensive if we compare it with others. But, when comparing to OMS, The Azure monitor is a basic one that we can use for resource metrics and activity logs.

Assume, if somebody stopped virtual machine that we want to get alerted or if the CPU utilization in one of the virtual machines is beyond 90 percent, then we want to get alerted. We can define these rules in Azure monitor. Once the condition within the rule is satisfied, then we can take a number of actions as a result of that. We can trigger Azure automation Runbook, azure function, logic app, or third party API.

Health Monitoring
Azure service health

❖ It provides timely and personalized information when problems in Azure services impact your services.
❖ It helps you prepare for upcoming planned maintenance.

Azure resource health

❖ Resource health helps you diagnose and get support when an Azure issue impacts your resources.
❖ It can be used to view the current and past health of your Azure resources.

❖ It provides technical support when you need help with Azure service issues.

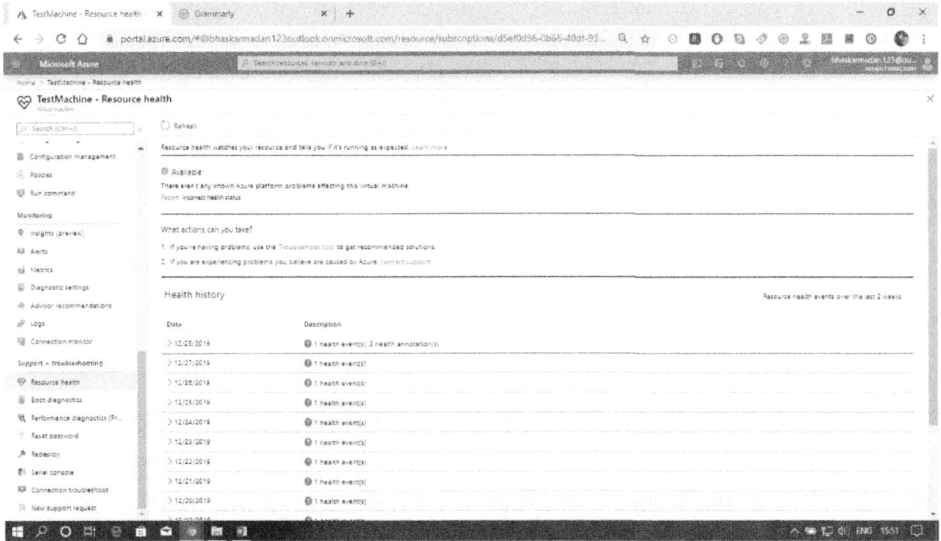

Advanced monitoring

❖ **Operations management suite (OMS):** It includes monitoring, alerting, and alert remediation capabilities across cloud and on-premises assets.

❖ **Log analytics:** It collects data generated by resources in our cloud and on-premises environments and from other monitoring tools to provide analysis across multiple resources.

❖ **Network Watcher:** It monitors our VM and its associated resources as they relate to the network that they are present.

Azure Cloud Service

Cloud Service is a Platform as a Service that is designed to support web applications that are scalable, reliable, and cheaper to operate. Using cloud service, we can deploy a web application into Azure. We have more control over Virtual Machines. We can install custom software on VMs that uses Azure Cloud Service, and we can access them remotely.

Using cloud service, we don't create virtual machines. Instead, we provide a configuration file that tells Azure how many instances we would like to create, the size of the instance, and the platform will create them for us.

Cloud service is able to detect any failed VMs and applications and ready to start new VMs or application instances when a failure occurs. Cloud service applications shouldn't maintain state in the file system of its own VMs.

Cloud Service Roles

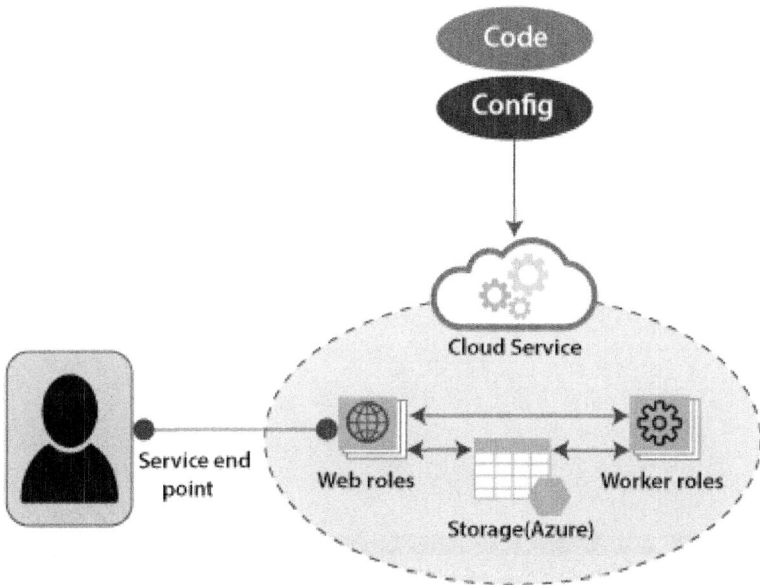

Web role: It automatically deploys and hosts our app through IIS.

Work role: It does not use IIS and runs our app standalone. If we want to run any continuous bathes, then we can use worker roles, and both the Web role and Worker role will interact with storage to get an application package, etc.

To deploy these Web roles and Worker roles, we will provide configuration and code associated with these web applications.

Cloud Service Components

Three key components constitute a cloud service.

- ❖ **ServiceDefenition.csdef** file specifies the settings that are used by Azure to configure the cloud service. For example - sites, endpoints, certificates, etc.
- ❖ **ServiceConfiguration.cscfg** contains the values that will be used to determine the configuration of settings for the cloud service. For example - number of instances, types of instances, ports, etc.
- ❖ **Service package.cspkg** used to deploy the application as a cloud service. First, it needs to be packaged using the CSPacK command-line tool. CSPacK generates an application package file that can be uploaded into Azure using the portal.

Creating cloud service using the Azure portal

Step 1: Click on create a resource and then type-in Cloud Service.

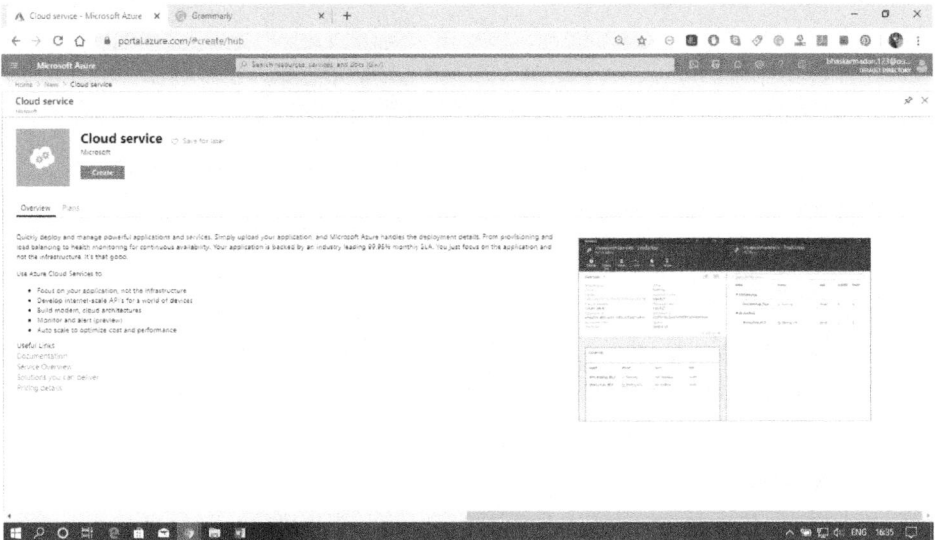

Step 2: After that, click on it and then click on create.

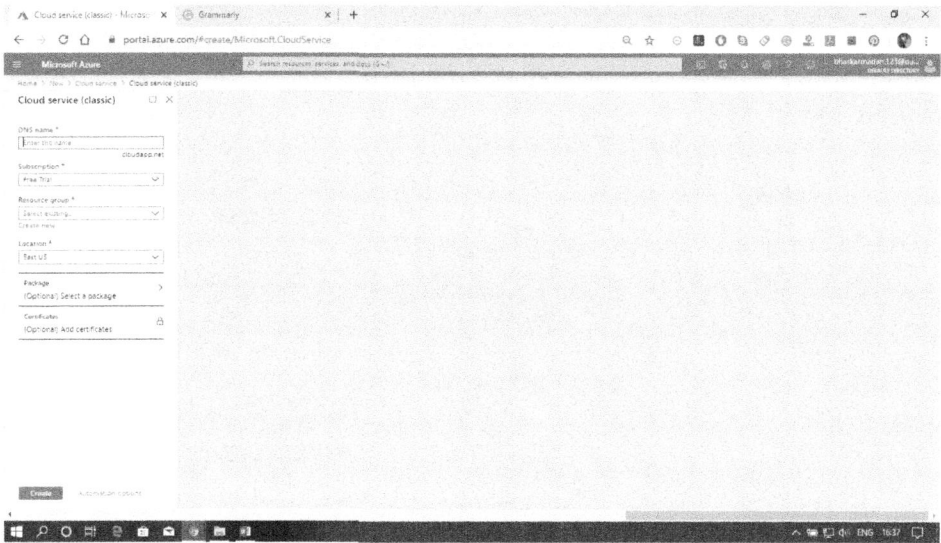

Step 3: Fill-in the DNS name, select the resource group, and location.

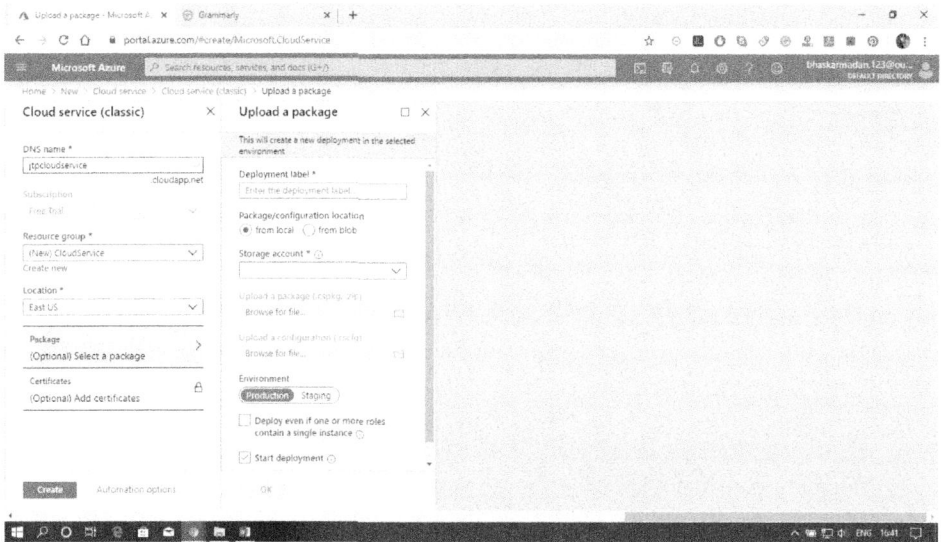

Step 4: Now, Click on create. Your cloud service will be created.

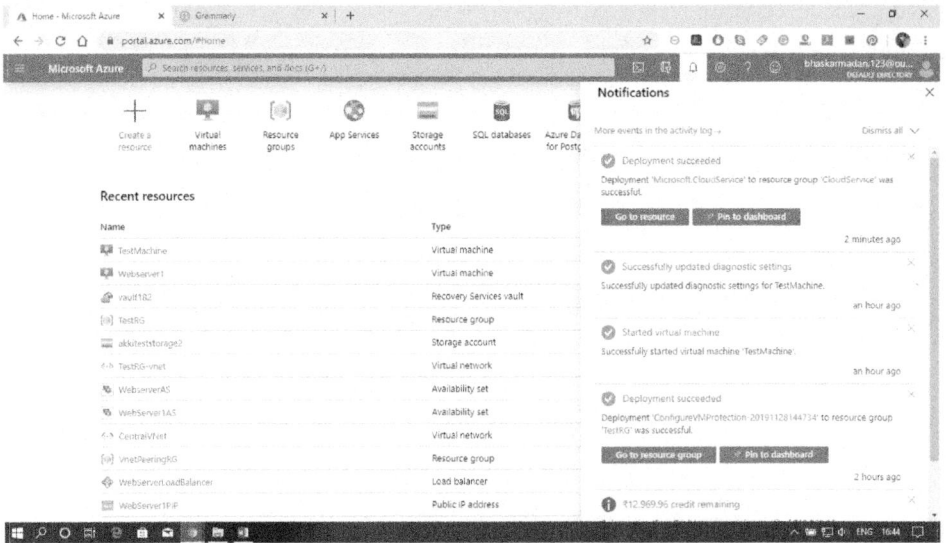

Step 5: To view the cloud service, click on go-to resources.

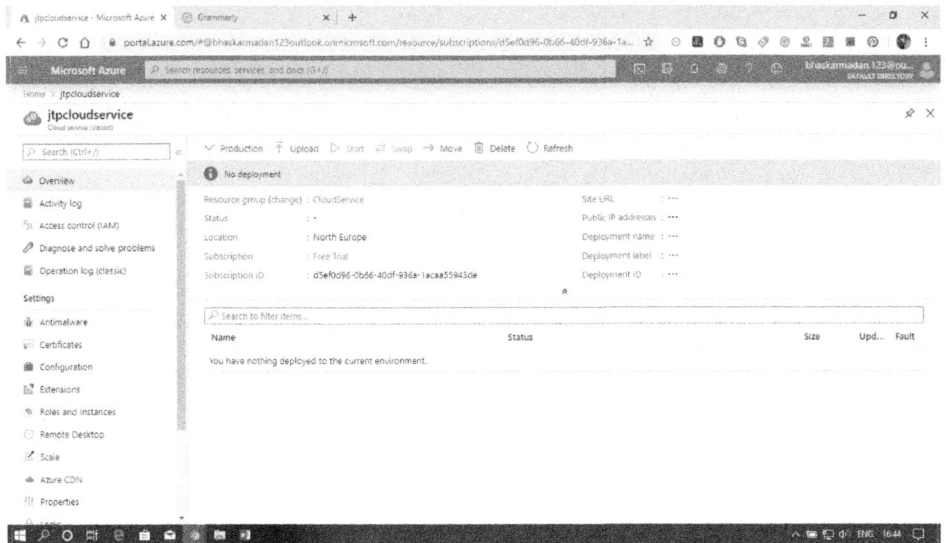

Step 6: Now, go to Visual Studio and create a new cloud service project. Here you can see the basic configuration setting, as shown in the image below.

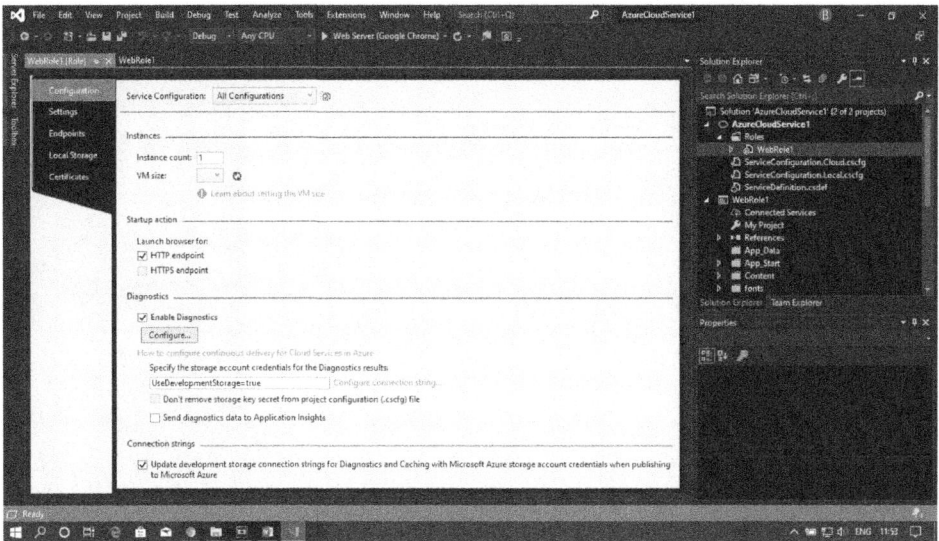

Step 7: To publish this cloud service into Azure, right-click on the file name. Then click on publish.

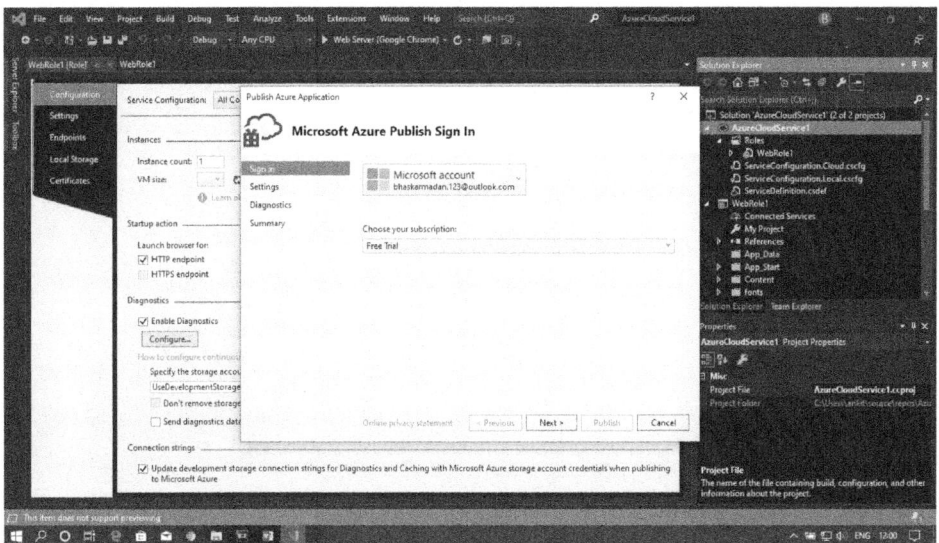

Step 8: Select your subscription and click next.

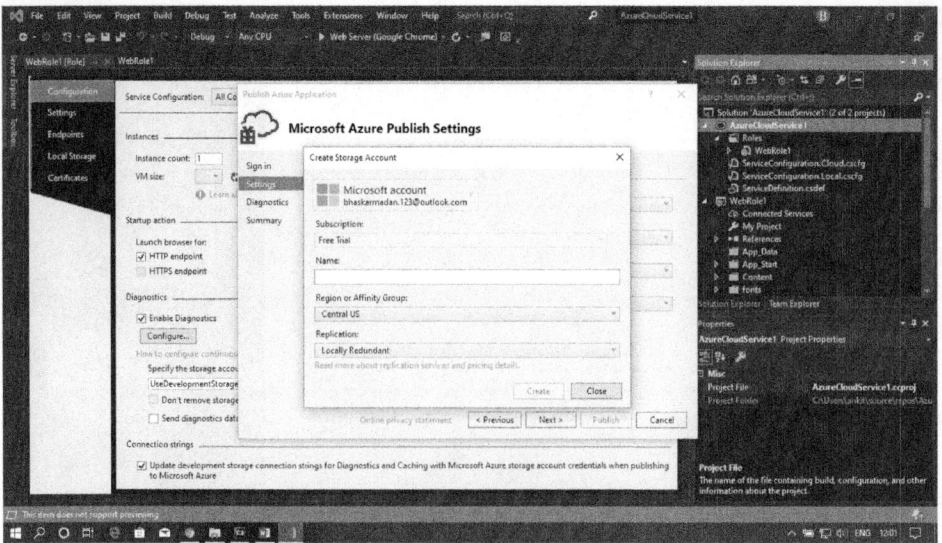

Step 9: Fill all the required details and then click on publish.

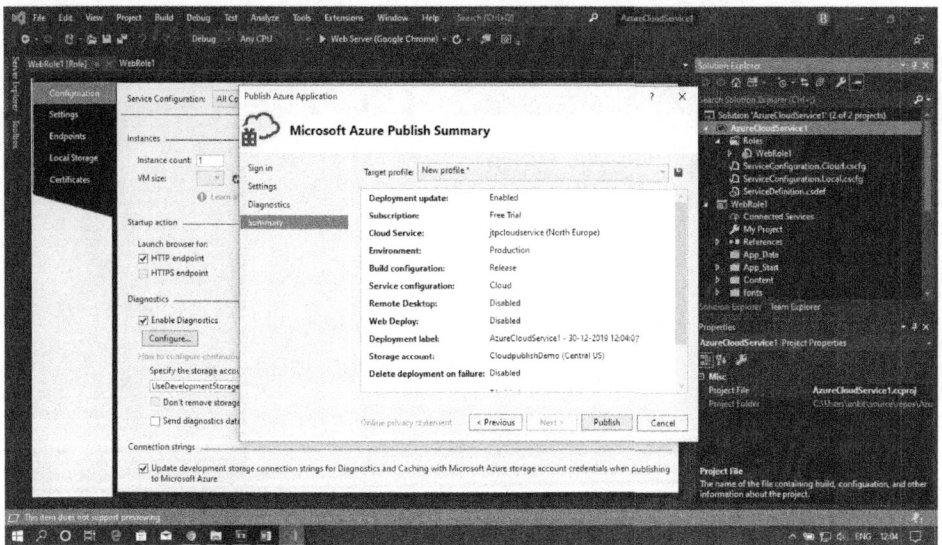

Step 10: Now, go to the Azure portal and click on the Resource group that you have created. You can now see your web role published, as shown in the figure below.

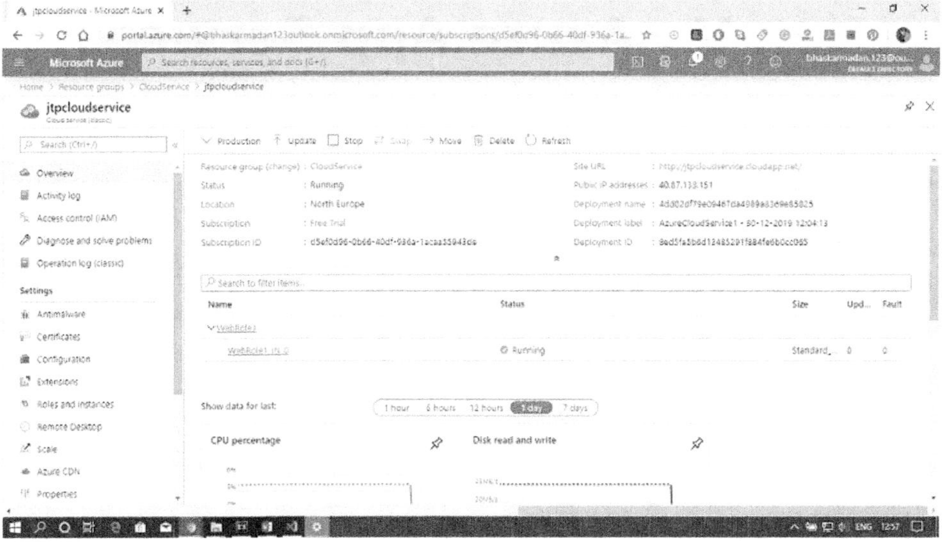
Microsoft Azure
jtpcloudservice
Production
Update
Stop
Move
Delete
Refresh
Overview
Activity log
Access control (IAM)
Diagnose and solve problems
Operation log (classic)
Settings
Antimalware
Certificates
Configuration
Extensions
Roles and instances
Remote Desktop
Scale
Azure CDN
Properties
Status
: Running
Location
: North Europe
Public IP addresses
: 40.87.133.151
Name
Status
Size
Running
Show data for last:
1 hour
6 hours
12 hours
1 day
7 days
CPU percentage
Disk read and write

Section V : Azure App Services

Azure App Services

The most fundamental building block of Azure App Service is the App Service plan or App Service environment.

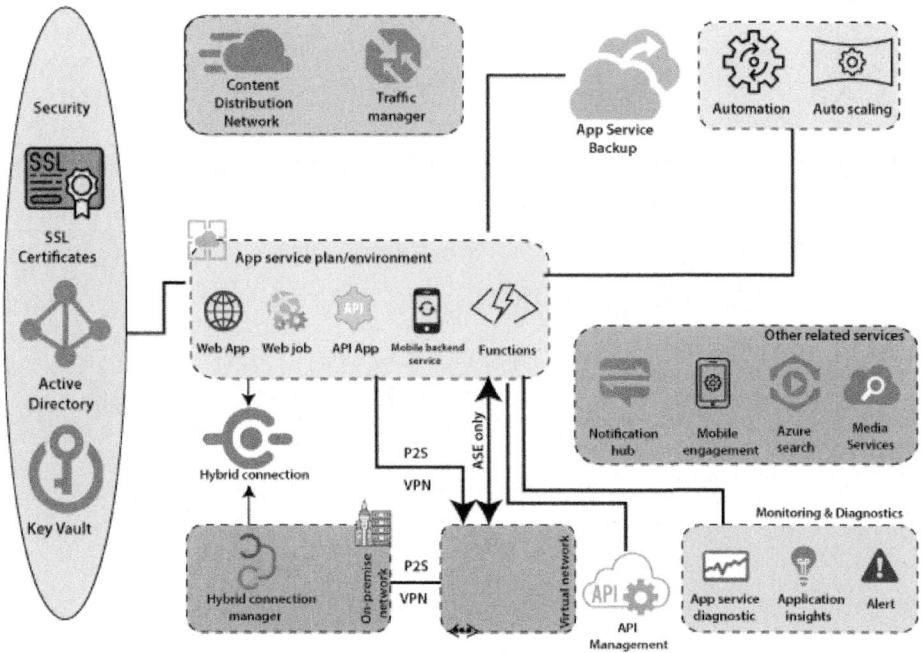

There are two types of hosting environments within App Service. App Service plan and App Service environment. App Service Environment is a more sophisticated version of the App Service plan and comes with a lot more features when compared to the App Service plan. Within these, we can host several Apps like - web applications, web jobs, batches, APIs, and mobile backend services that can be consumed from our mobile Front-End.

Other related services are closely related to these apps within the App service plan. Those related services are a notification hub that we can use to push notifications into mobile devices. We can use Mobile engagement to carry out Mobile analytics.

Apart from these related services, there is one more service, which is very important when it comes to APIs, which is API management. API management can act as a wrapper around our API apps when we're exposing those APIs to the outside world. It comes with a lot of features such as throttling, security, and it will be beneficial if we want to commoditize our APIs and sell it to the outside world.

To enable communication between apps in the App Service plan and apps installed on virtual machines within the virtual network. There are two ways we can do it. One way is to establish Point-to-site VPN between apps in the App Service plan and virtual network via which the apps can communicate with each other. And the second way is if we have the App service environment. Because it will get deployed into a virtual machine by itself, the Apps within that App Service environment can seamlessly communicate with the apps installed on virtual machines within the virtual network.

And finally, there are two important things. The first one is security, and the second one is monitoring to secure and control the App services environment.

App Service plan

An app service plan denotes a set of features and capacity that we can share across multiple apps in the same subscription and geographical region. A single or dual app can be configured to run on the same computing resources.

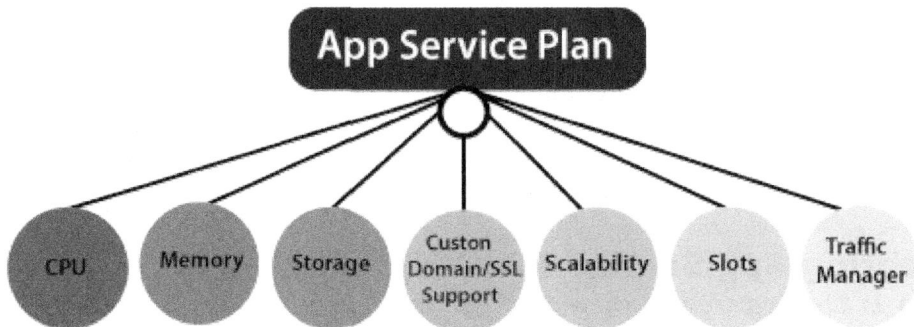

Each App Service plan defines:

- ❖ Region (West US, East US, etc.)
- ❖ Number of VM instances
- ❖ Size of VM instances (Small, Medium, Large)
- ❖ Pricing tier

❖ *Shared compute:* Free and shared, the two basic tiers, runs an app over the same Azure VM as other App Service app runs, including apps of different customers.

❖ *Dedicated compute:* Basic, Standard, Premium, and PremiumV2 tiers run apps on a fixed Azure VM.

❖ *Isolated:* This tier runs dedicated Azure VMs on dedicated Azure Virtual Networks, which provides network isolation on top of computing isolation to your apps.

❖ *Consumption:* It is only available to function apps. It scales the functions dynamically, depending on the workload.

Environment features

❖ Development frameworks: App Service supports a variety of development frameworks, including ASP.NET, classic ASP, node.js, PHP, and Python- all of which run as extensions within IIS.

❖ File access
   ❖ *Local drives* - Operating system drive (D:\drive), an application drive and user drive (the C:\ drive)
   ❖ *Network drives* - Each customer's subscription has a reserved directory structure on a specific UNC share within a data center.

❖ Network access: The application code can use TCP/IP and UDP based protocols to make outbound network connections to access Internet endpoints that expose external services.

Web apps Overview

Azure App Service Web Apps is a service for hosting web applications. The key feature of App Service Web Apps.

❖ Multiple language and frameworks
❖ DevOps optimizations
❖ Security & Compliance
❖ Application template
❖ Visual Studio integration

Creating App Service Plan in Azure Portal

Step 1: Click on *create a new resource* and search for App Service Plan to create it.

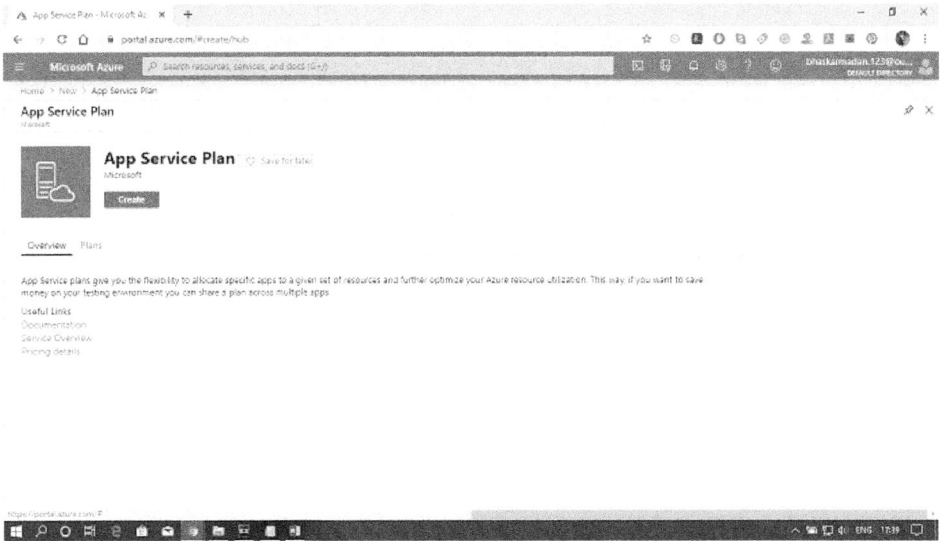

Step 2: Fill-in all the required details and select the SKU size, as shown in the figure below. Then click on create.

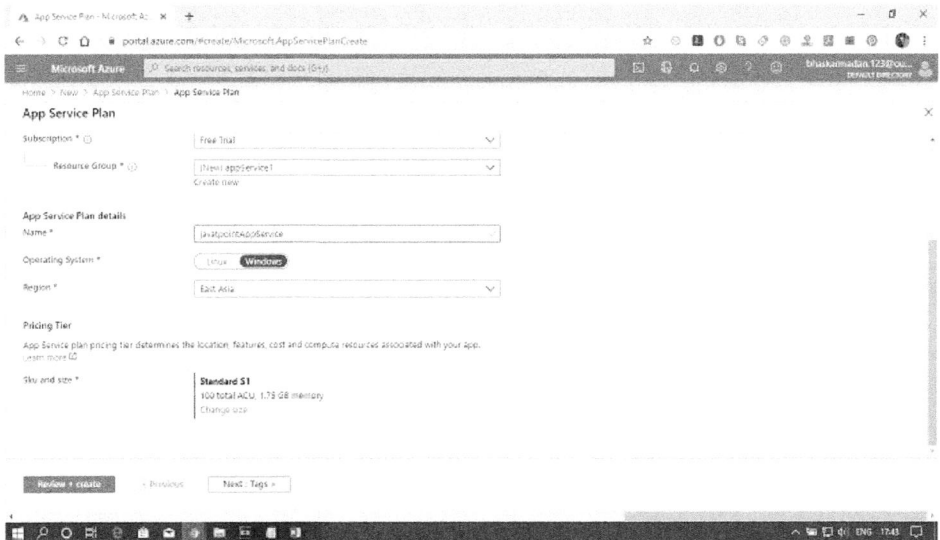

Step 3: Your app service plan will be created. You can now explore and modify it as per your requirement.

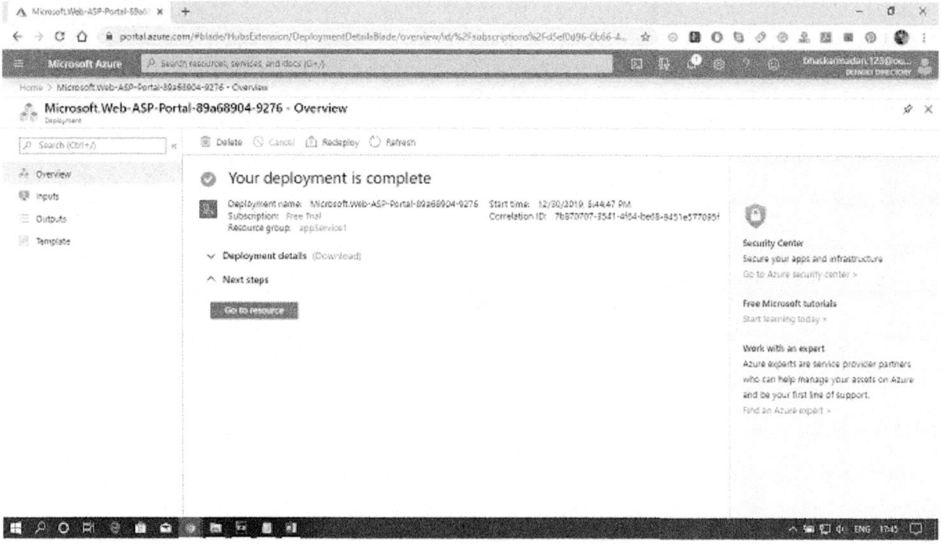
Microsoft Azure
Home > Microsoft.Web-ASP-Portal-89a68904-9276 - Overview
Microsoft.Web-ASP-Portal-89a68904-9276 - Overview
Deployment
Search (Ctrl+/)
Delete
Cancel
Redeploy
Refresh
Overview
Inputs
Outputs
Template
Your deployment is complete
Deployment name: Microsoft.Web-ASP-Portal-89a68904-9276
Subscription: Free Trial
Resource group: appService1
Start time: 12/30/2019, 5:44:47 PM
Correlation ID:
Deployment details (Download)
Next steps
Go to resource
Security Center
Secure your apps and infrastructure
Go to Azure security center >
Free Microsoft tutorials
Start learning today >
Work with an expert
Azure experts are service provider partners who can help manage your assets on Azure and be your first line of support.
Find an Azure expert >

Azure Web App

Azure Web App service lets us quickly build, deploy, and scale enterprise-grade web, mobile, and API apps running on any platform. It helps us to meet rigorous performance, scalability, security, and compliance requirements while using a fully managed platform to perform infrastructure maintenance.

Creating a Web App and deploying an application into Azure web App from visual studio

Step 1: Click on create a resource and type in the web app. After that, click on the web app and then click on Create.

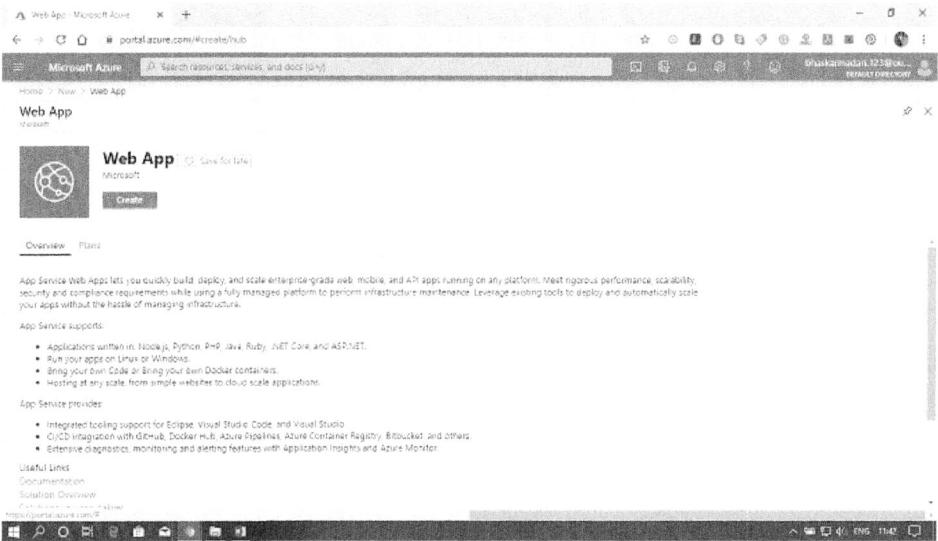

Step 2: You are now on the Web App creation page. Fill-in, all the required details, then click on review+create.

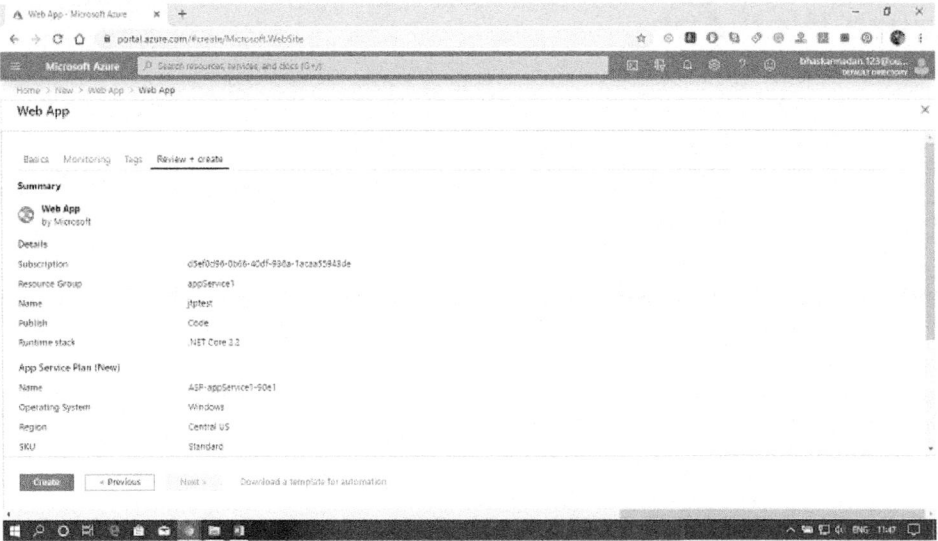

Step 3: Click on create, then you will be redirected to the following page.

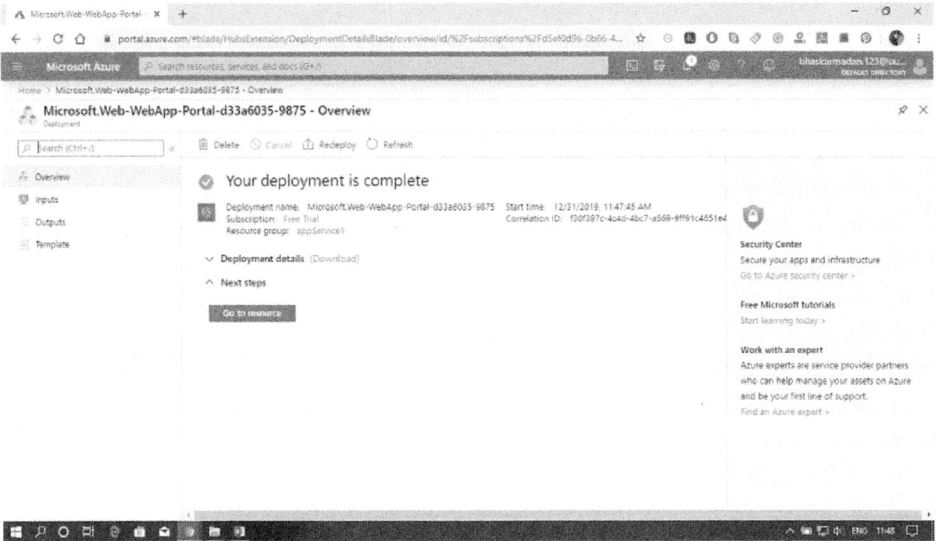

Step 4: Open Visual Studio, then click on Create a new project.

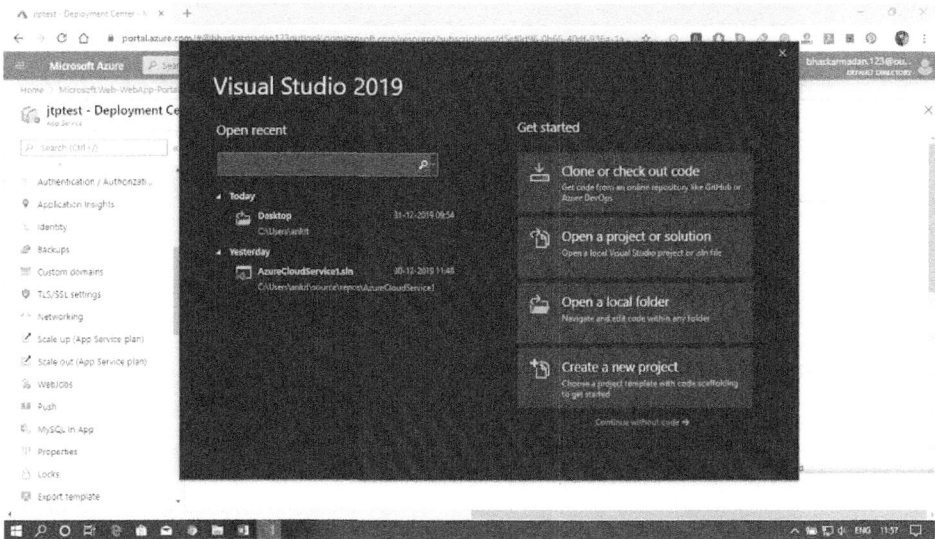

Step 5: Search for ASP.net web application and click on it.

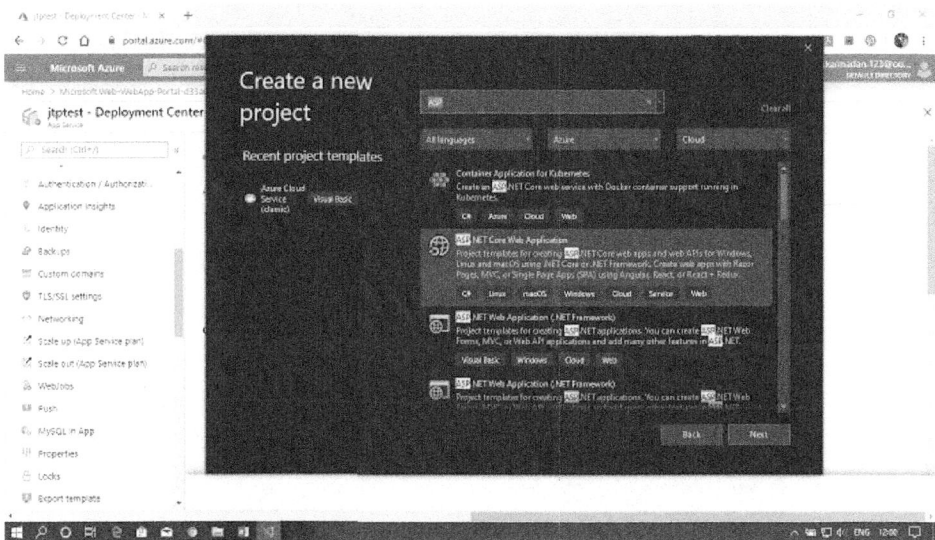

Step 6: Now, configure your project and click on create.

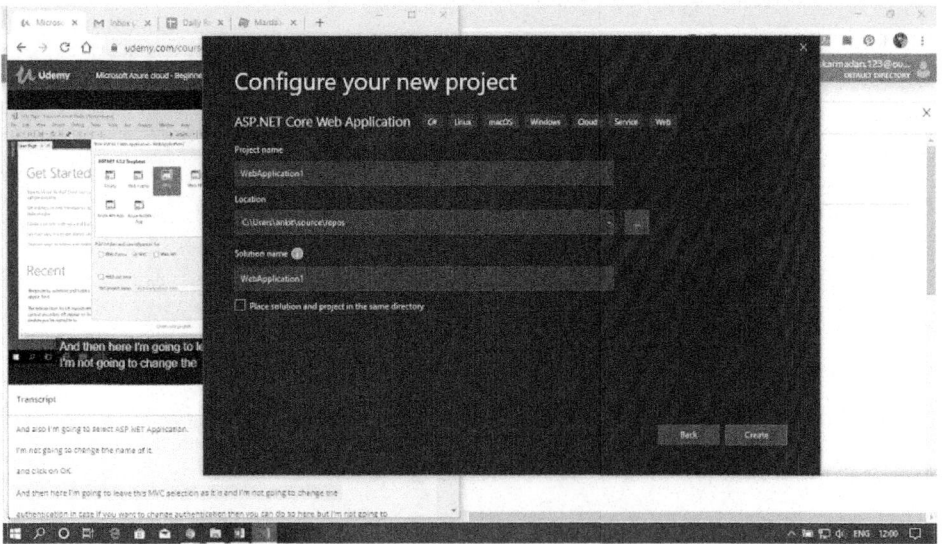

Step 7: Now select the Web application (Model View Controller) option from available templates. Then Click on create.

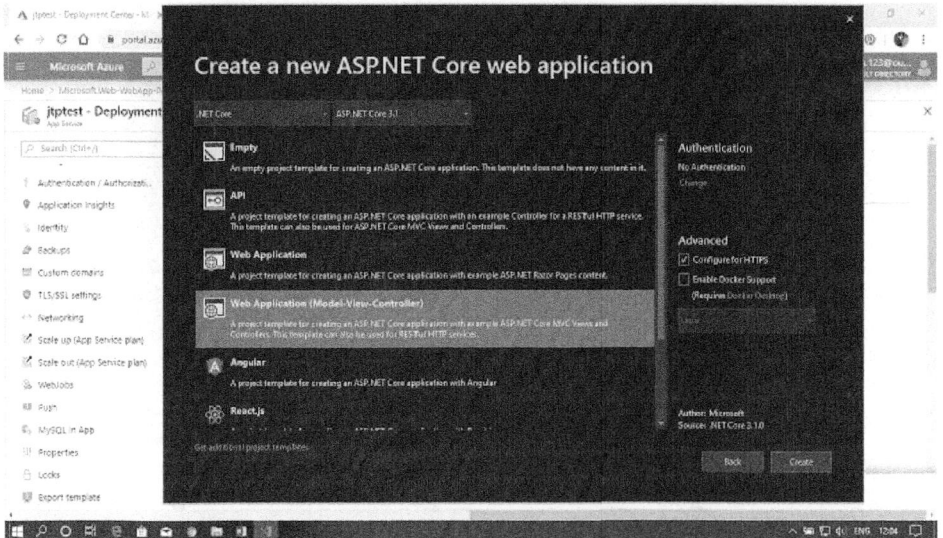

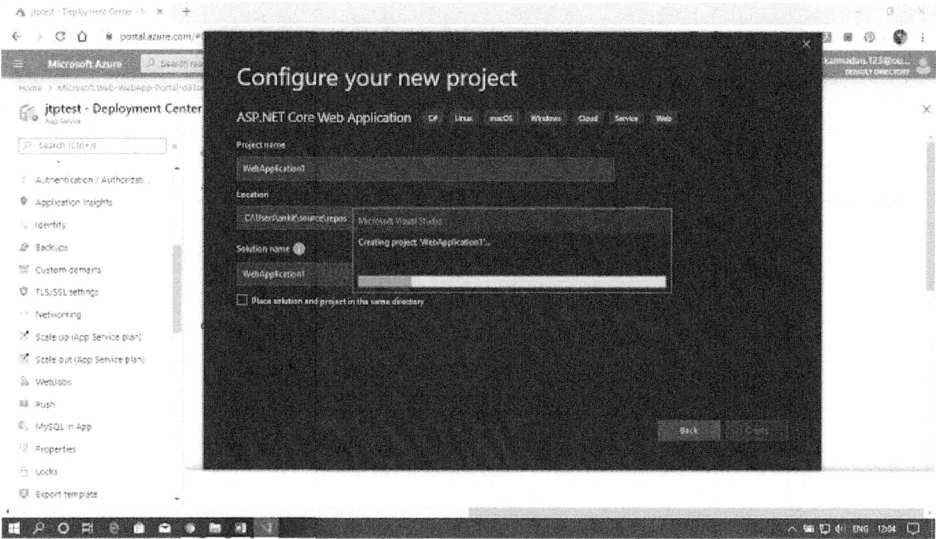

Step 8: Your project will be created, Now click on publish to configure it with the Azure portal.

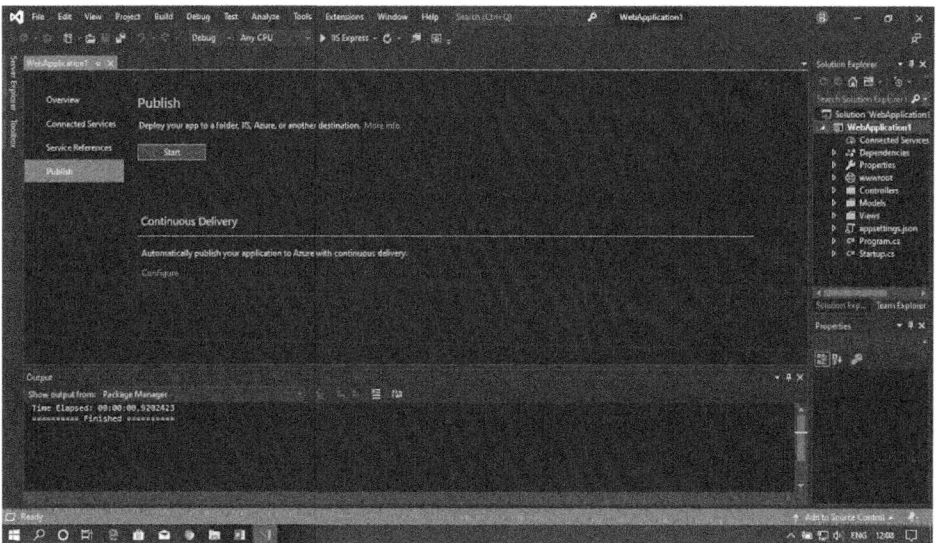

Step 9: Here, either you can create a new service plan, or you can use an existing one.

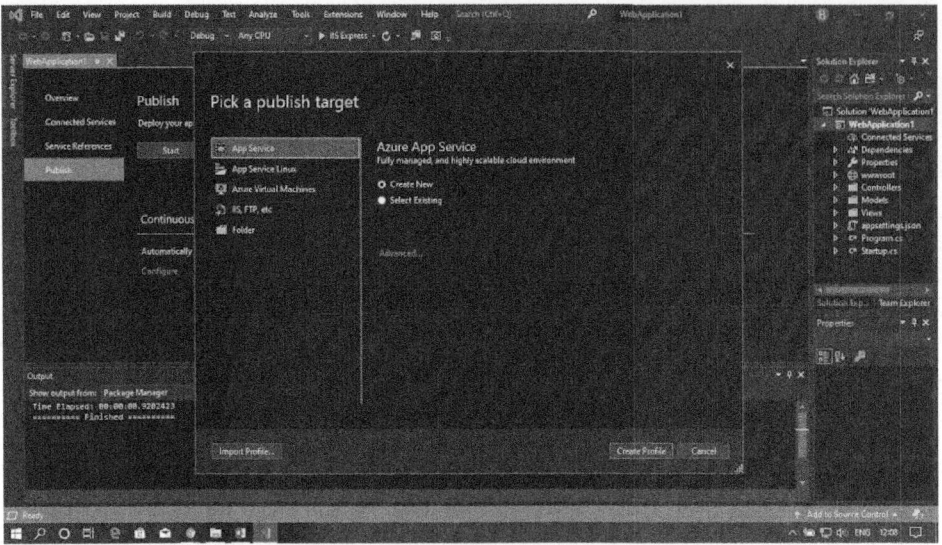

Step 10: Let's see how to create a new one, click on Create new. Fill all the details and click on create a profile.

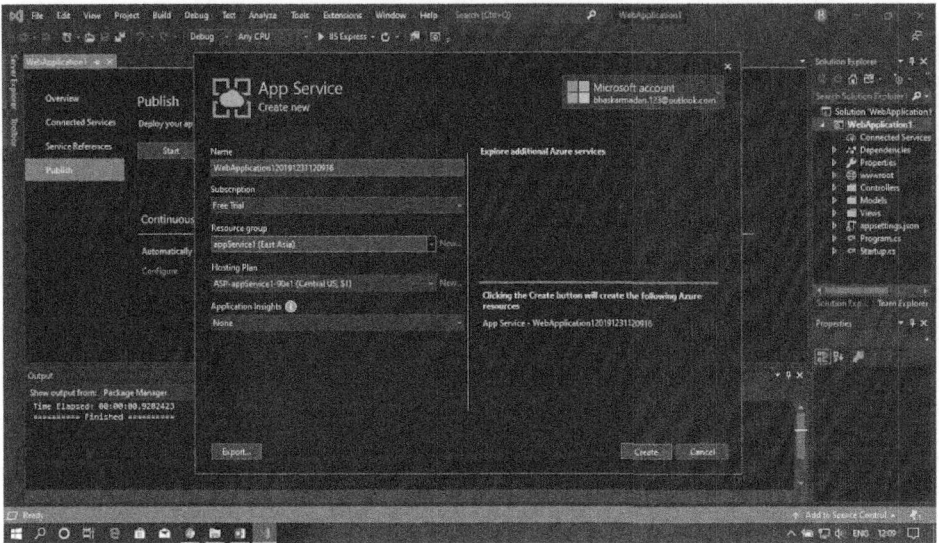

Step 11: As we have already created an App service previously, so we are using that one here. Go back and click on *Select existing*.

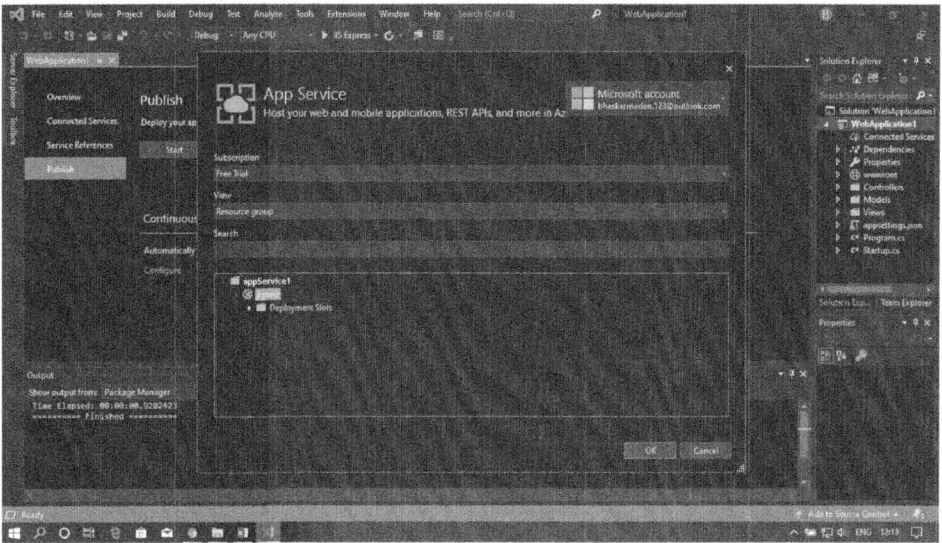

Step 12: Now, click on the file name and then click on, Ok.

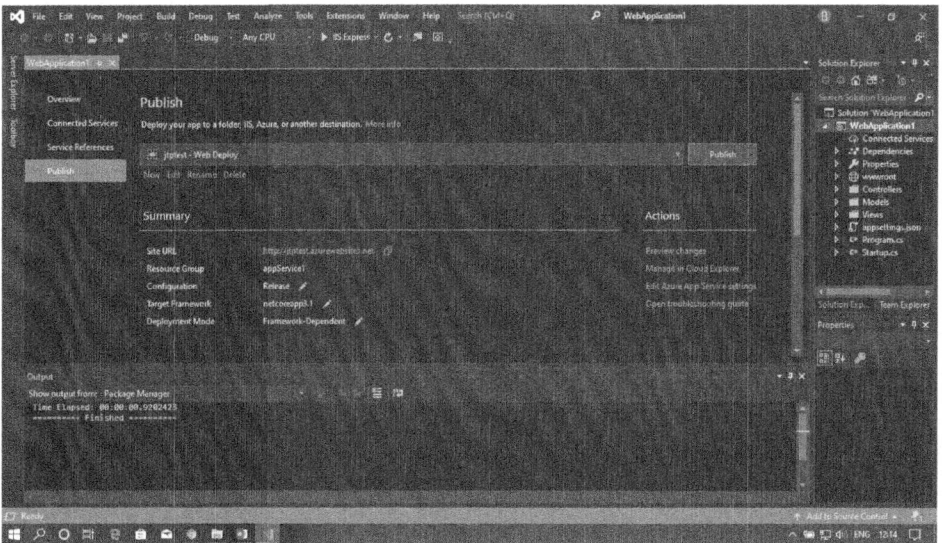

Step 13: Now go to the Azure portal and click on the storage account.

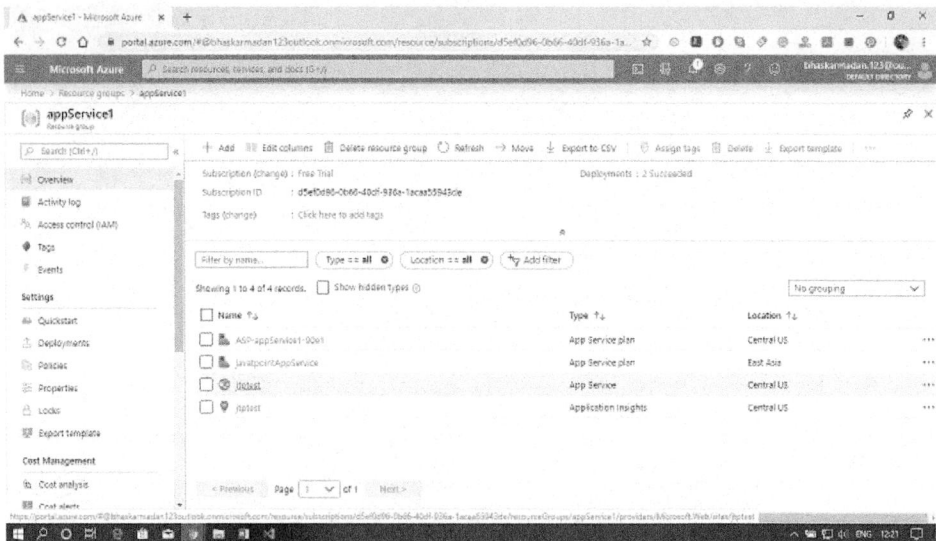

Step 14: Click on the app section, here you can view the web app that you have created.

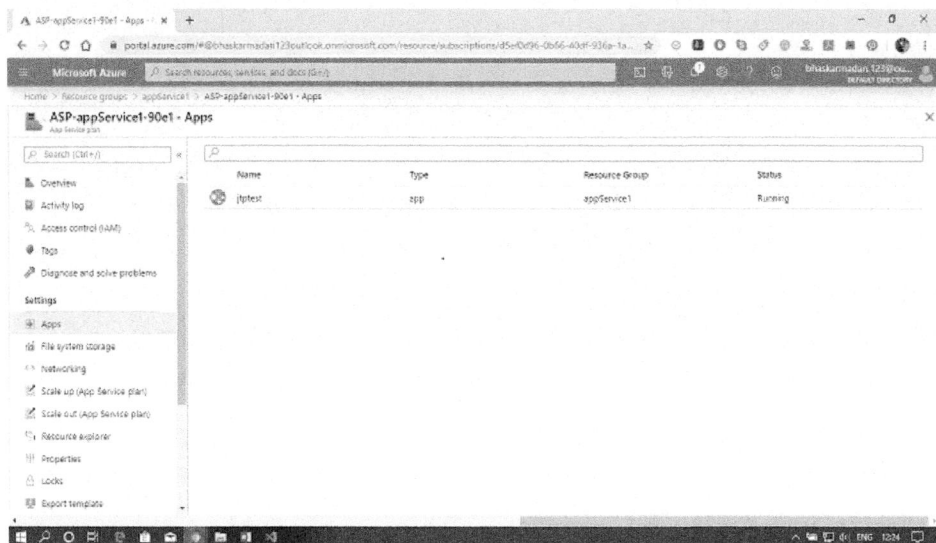

Step 15: Click on the browse button above to see that your Web app is working or not.

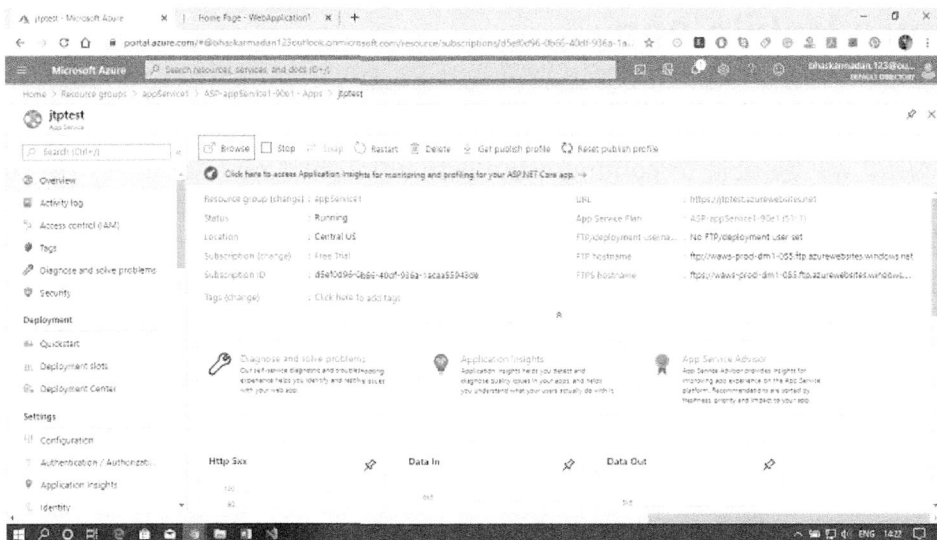

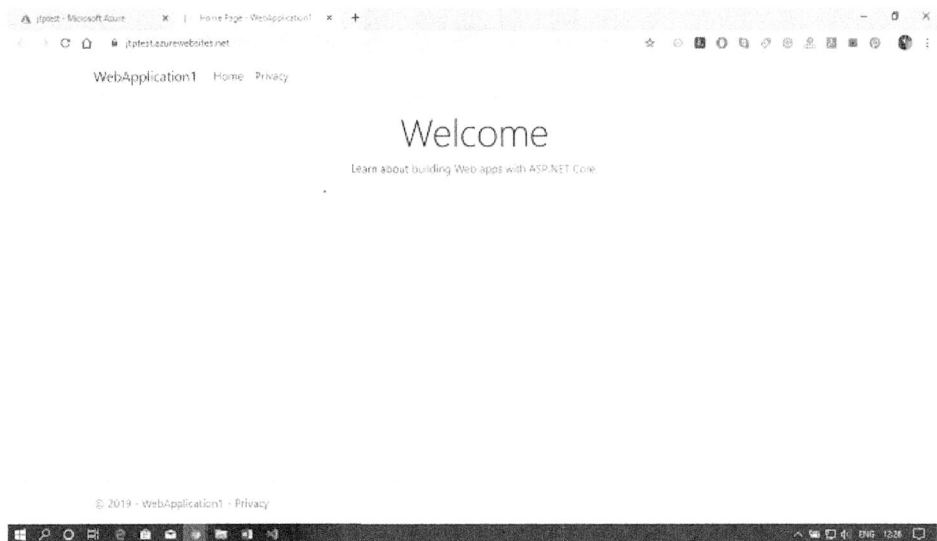

Azure Mobile App

We can deploy our mobile backend services on Azure using Azure Mobile apps. By implementing our mobile backend service on Azure, our mobile backend will be able to communicate with different Azure services. We can able to take advantage of various features that are provided by Azure Mobile Apps.

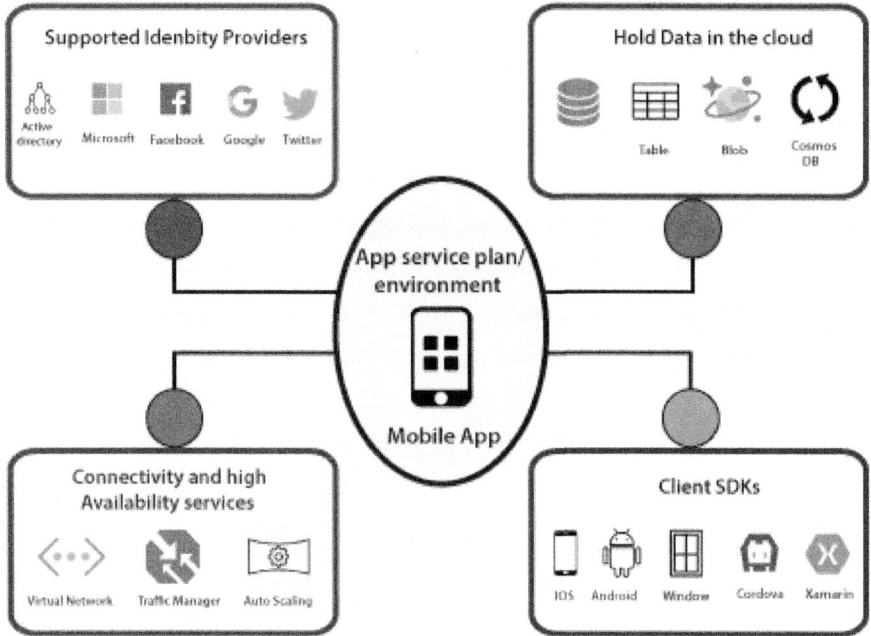

Features and services

Data Storage: Our mobile backend can be able to store the data or access the data of Azure SQL database Azure table storage, Azure Blob storage, and Cosmos DB. And also, we can add offline sync features to our mobile app.

Client-Side development: Once we host our mobile backend services on Azure, there must be a way to consume those services. For that purpose, Microsoft provided several client SDK's depending upon the platform.

Authentication and Authorization: We can integrate our mobile app with different authentication providers. So we can integrate with Azure active directory, Microsoft, Facebook, Google, and Twitter. We can integrate this service without any code.

Connectivity: In terms of connectivity to apps that are hosted in a virtual network. Our mobile app will be able to connect to a virtual network either using point to site VPN or by hosting our mobile app backend service into an app service environment, which will automatically get connected to the virtual network.

Availability Service: We can use the traffic manager to provide higher resilience even during the data center failures. We can also use auto-scaling to scale our mobile backend services as more number of users are about to start using our app.

Notification Hub: Using the notification hub, we can enable push notifications to different platforms, and also we can use mobile engagement using which we can understand what exactly the user is doing with our app.

How to create Mobile apps backend development

Step 1: Log in to the Azure portal and create a new Azure mobile app backend.

Step 2: Configure the mobile app backend.

Step 3: Define a table controller.

Step 4: Create the Data Transfer Object (DTO) class.

Step 5: Configure a table reference in the Mobile DbContext class.

Step 6: Create a table controller.

Step 7: Define a custom API controller.

Mobile Client-side development

Step 1: Based on mobile OS, download the client-side SDK.

Step 2: Reference the MicrosoftAzureMobile (IOS) in your client code.

Step 3: Create MSClient (IOS) reference and start accessing data from tables.

Step 4: For Custom API's, use MSClient.invokeAPI to call custom API.

Mobile offline data sync

Mobile offline data sync is a client and server SDK feature of Azure Mobile Apps that makes it possible to create apps that work without a network connection.

Sync Table

❖ To access the "/tables" endpoint, Azure Mobile client SDKs provide an interface such as MSTable. However, this will fail if the client devices do not have a network connection.

❖ To support offline use, our app should instead use the sync table APIs such as MSSyncTable. All the CRUD operations will happen at a local store.

The local store is the data persistence layer on the client device (Windows, Xamarin, and Android). It is based on SQLite, whereas on iOS, it is based on core data. In offline synchronization, the sync can be a push, pull, implicit pushes, or Incremental sync.

Azure Notification Hub & Mobile Engagement

Azure notification hub provides an easy-to-use, multiplatform, scaled-out push infrastructure that enables us to send mobile push notification requirements in which you want to send notifications to the users. Using the Azure notification hub, we can achieve the same with minimum code and minimum configuration. With a single API call, we can target individual users or entire audience segments containing millions of users across all the devices.

Azure notification hub implements all the functionality of push infrastructure. The only thing we need to do is to write a mobile app in such a way that the mobile app will register the PNS handle with Azure Notification, and our mobile backend will be responsible for sending platform-independent messages to all users and interest groups.

Advantages of Azure Notification Hub

Multiplatform: We can use Azure Notification Hub to send push notifications to IOS devices, Android devices, and Windows devices. And it works with any backend that is developed in any language.

Scalability: We don?t need to worry about scaling. Azure notification hub will take care of that for us. We can scale up to millions without changing anything.

Delivery pattern: We have a vibrant set of delivery patterns. We can broadcast, unicast, or multicast user segmentations. So we can divide all our users into segments, and we can target a specific part of users to send notifications using Azure Notification Hub.

Working of Azure Notification Hubs

Let?s understand the working of Azure Notification Hubs using a simple diagram.

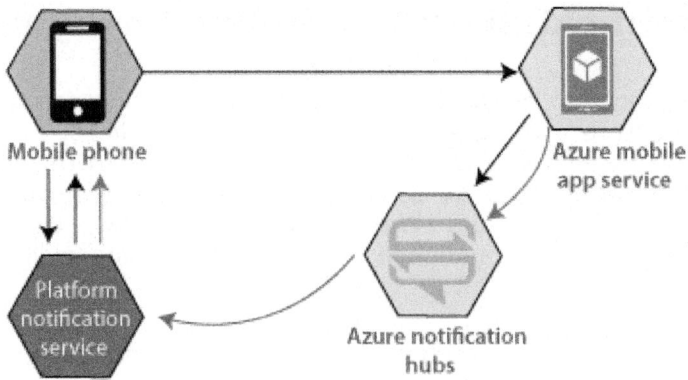

Firstly we need to do is to retrieve the PNS handle from the platform notification service so that our mobile client will retrieve the PNS handle and pass on that handle to Azure Notification Hub via our Azure mobile app backend service. It can be anything. And from that point onwards, our mobile backend service can interact with Azure notification hub to send notifications.

Mobile Engagement

Azure Mobile Engagement is a software as a service user Engagement platform that provides data-driven insights into app usage, real-time user segmentation. And the key thing is it enables contextually-aware push notifications and in-app messaging.

For example, ? We have an e-commerce website, and some of the users are showing more interest in sports-related equipment. In that case, using Mobile engagement, we can identify those users that are visiting the sports product more frequently. And whether we want to offer discounts or any new sports product that came into the market, then we can send a notification to only those users that are showing interest in sports products.

Another thing that we can do using Azure Mobile Engagement is data-driven insights into app usage. You can see which screen of your app get more engagement from the users using which we can improve our app. We can do all real-time user segmentation based on the user data and also based on the pages they visit, the type of data they search, etc.

The Azure Mobile Engagement can be used with Azure mobile apps, which makes in total a compelling platform for the development of your mobile apps.

Azure API Apps and API Management

The API apps features make it easy to develop, host, and consume APIs in the cloud and on-premises. The advantage of hosting APIs in Azure API apps is that we will get enterprise-grade security and simple access control, automatic SDK generation, and seamless integration with Logic Apps. Logic Apps are system workflows that you can build within Azure. And as a part of the workflow, each activity needs to interact with the functionality exposed by a different system. By having those interfaces hosted in Azure, it makes it easy to integrate with the logic apps also.

Features of API apps

❖ **Bring our own existing API as-is:** API can be developed in any language framework supported by App Service such as C#, Java, PHP, Node.js, etc.

❖ **Easy Consumption:** There is integrated support for Swagger API. By enabling swagger, we are making it easy for others to consume our APIs, and also we will provide excellent visibility of APIs to developers.

❖ **Simple access control:** Protect an API app from unauthenticated access with no changes to your code.

❖ **Visual Studio Integration**

❖ **Integration with Logic apps**

API Management

❖ API Management is all about managing APIs. We can put an API Management frontend on an API to monitor and throttle usage, manipulate input and output, consolidate several APIs into one endpoint, and so forth. The APIs being managed will be hosted anywhere.

❖ API Apps is about hosting APIs, whereas API management is about managing APIs. Let?s see how API management works.

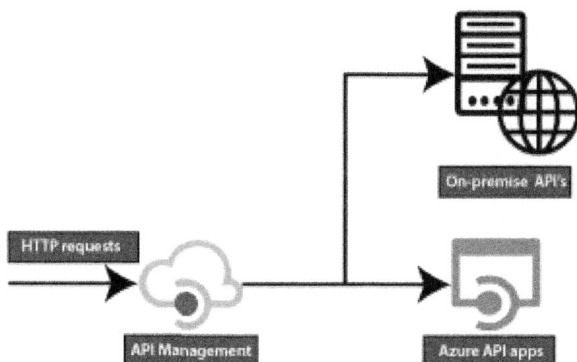

At a very high level, firstly, when http or https request comes. It will come to API management and the API management based on the location of the API. Then it will forward that request to either Azure API apps or on-premises apps. But when it is forwarding that request it can throttle, it can also monitor and manipulate the inputs and outputs.

API Management portals

❖ The API management portal is where developers can learn about APIs, view and call operations, and subscribe to products.

❖ Content within the developer portal is modified via the publisher portal, which is accessible from the Azure portal. To reach there, click on the Publisher portal from the service toolbar of our API Management instance.

❖ The dashboard of the developer portal can be customized by adding custom content, customizing styles, and adding our branding.

API management concepts

The API management concept is the crucial thing that we need to remember.

❖ **APIs and operations:** Each API represents a set of actions available (might be CRUD operation) to developers.

❖ **Products:** This is how APIs are surfaced to developers. Each product can contain multiple APIs.

❖ **Groups:** It is used to manage the visibility of APIs so we can have three types of groups.

❖ An administrator group member can manage API management service instances, creating the APIs, operations, and products that are used by developers.

❖ Developers? group members are authenticated customers that build applications using APIs.

❖ Guests are the unauthenticated developer portal users. Guests are our prospective customers who will come and consume/trail. They will view the APIs and see whether it fits into their requirements or not.

❖ **Policies:** It is a very powerful capability of API management that allows the publisher to change the behavior of the API through configuration, such as throughput.

Creating an API using Azure Portal

Step 1: Click on create a resource. After that type in API apps and click on create.

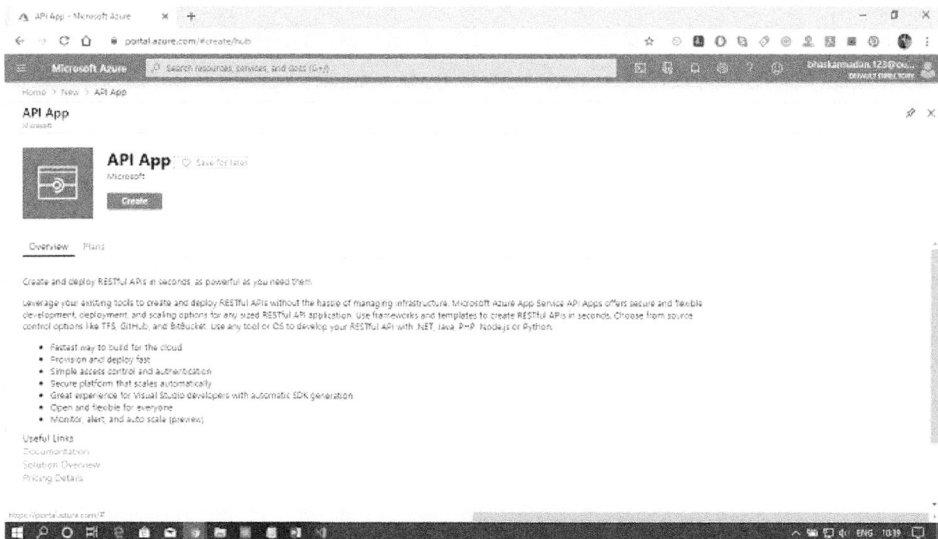

Step 2: Now, assign a name to your API app and select the resource group. After that, select the service architecture according to your requirements.

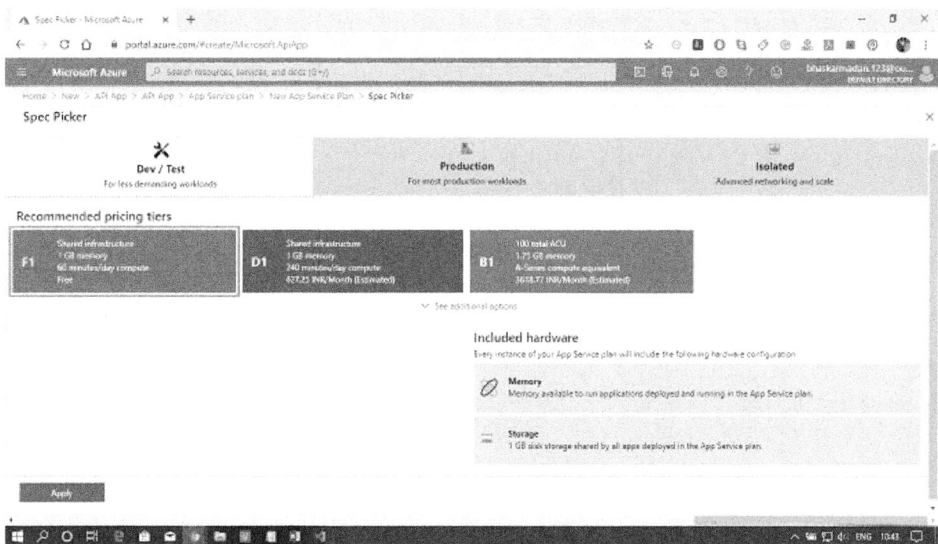

Step 3: Finally, click on create.

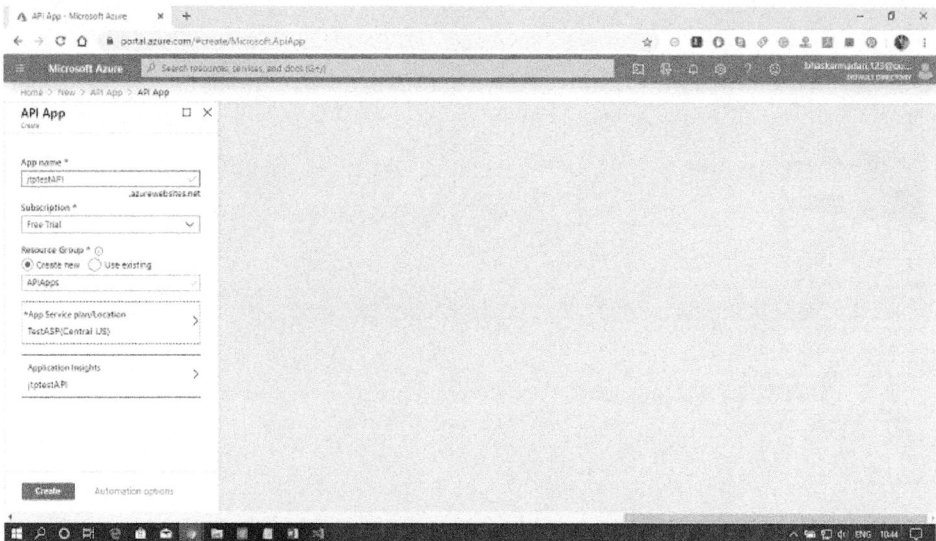

Step 4: Your API app will be successfully created.

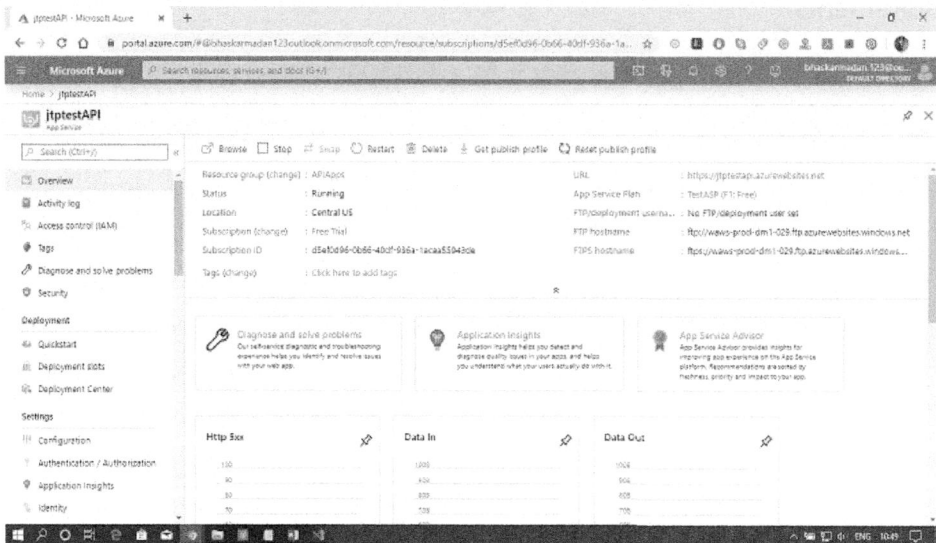

Publishing an API using Visual Studio

Step 1: Create a new web app project in Visual Studio. As shown in the following figure.

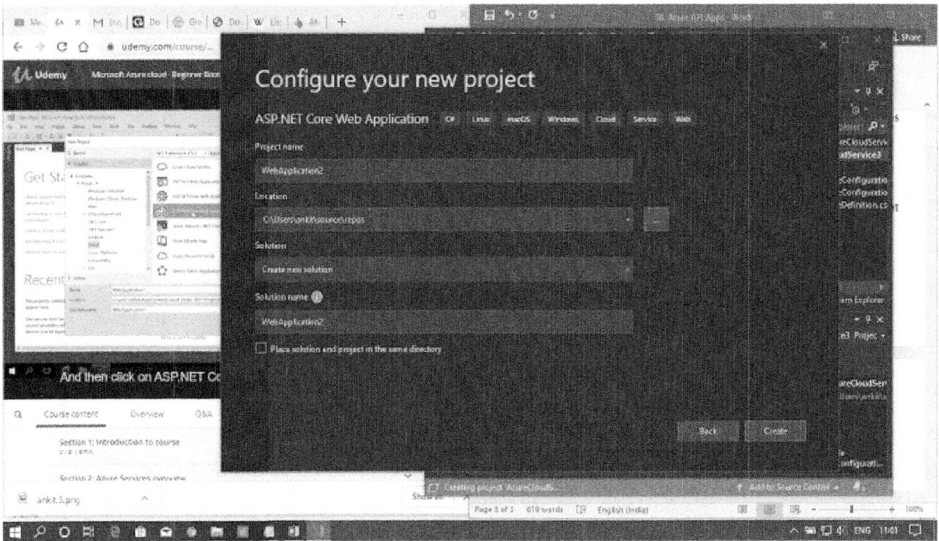

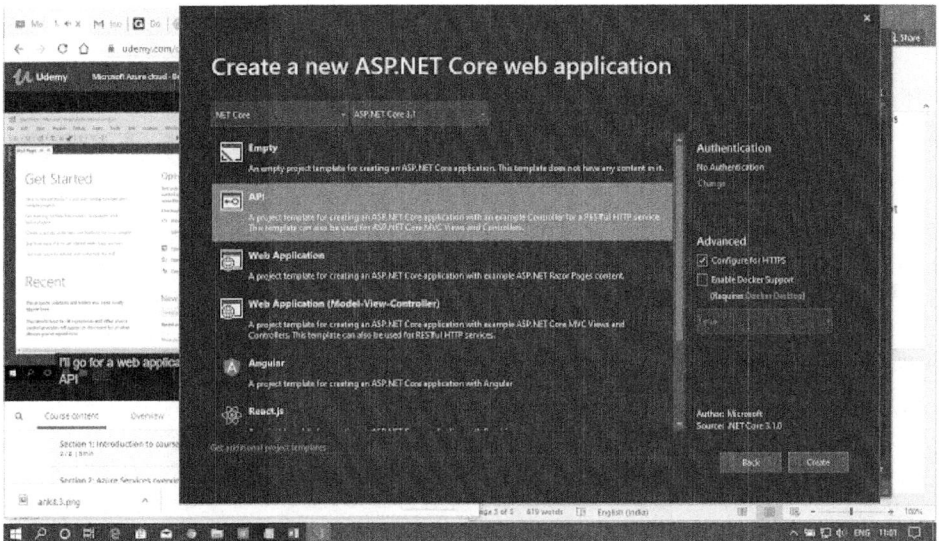

Step 2: Click on publish, then click on select existing. After that, click on the publish.

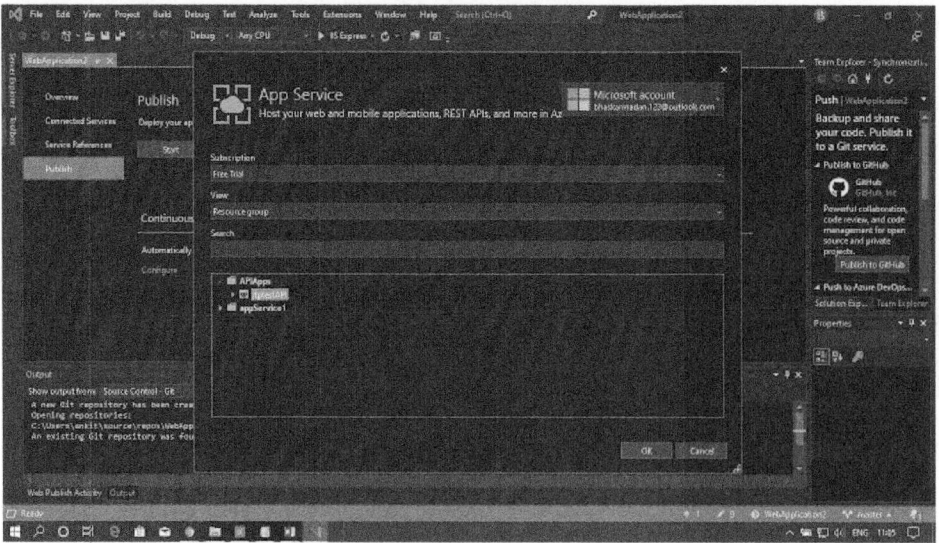

Step 3: Your web app is successfully posted to the Azure portal.

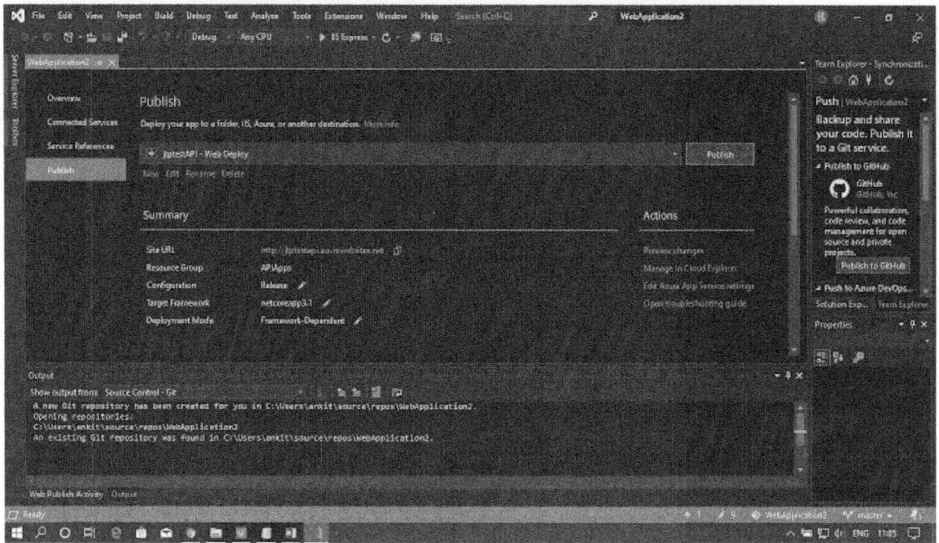

Azure App Service Backup

The App service backup and scaling is much simpler than virtual machine backup and scaling. The backup and restore in-app feature service let us quickly create app backups. This backup of app service will contain app configuration also, the file content, and optionally the database connected to our app. We can take backup along with the app service. The App service will have the following backup information:

* App configuration
* File content
* Database connected to our app

There are several ways we can take backup:

* Manually
* Automation based on scheduling
* Partial backup

The backup will be stored in a storage account. And in terms of restoration, we can restore an app with its linked database on-demand to its previous state, using the backup, or we can create all-together a new app using that app backup. Both backup and its restoration are only available for apps running in standard and premium tiers.

Scaling

There are two ways we can scale the app services.

Scale-up: It means we can get more CPU, memory, disc space, and also an extra feature like dedicated virtual machines, custom domains, certificates, staging slots, auto-scaling, and many other features based on the pricing tier we select when we are scaling up our app service plan.

Scale-out: It means we will increase the number of VM instances that run our app so we can scale out to any number of instances based on the pricing tier. But, if we go for app service environments in an isolated tier, then we can scale out to a hundred instances.

Apart from this, another important thing that we need to remember about scaling is Auto Scaling. There are many ways that we can scale our app services.

* Automatically
* Manually

- ❖ Pre-set Matric
- ❖ Scheduled

Scaling the App Service using Azure Portal

Step 1: Open your already created app service or create a new one.

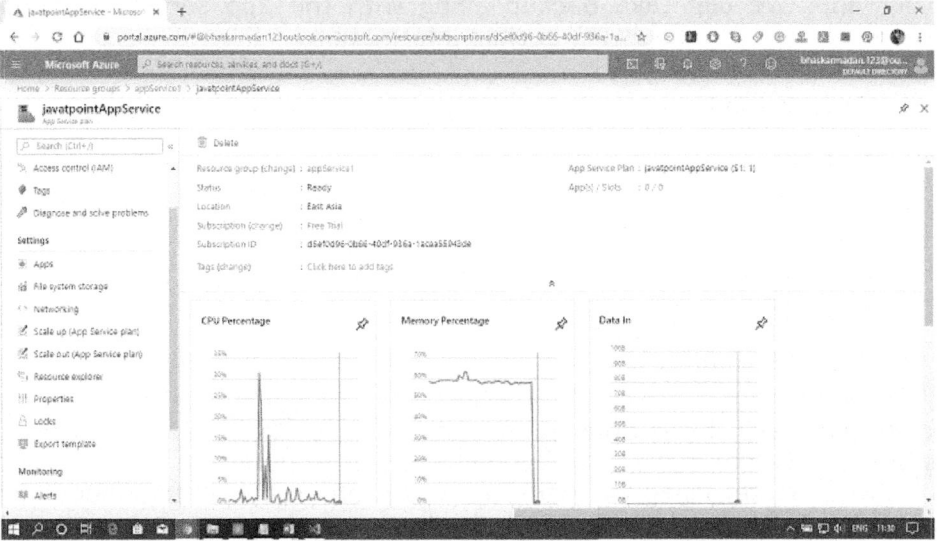

Step 2: Now, click on scale-up on the left toolbar.

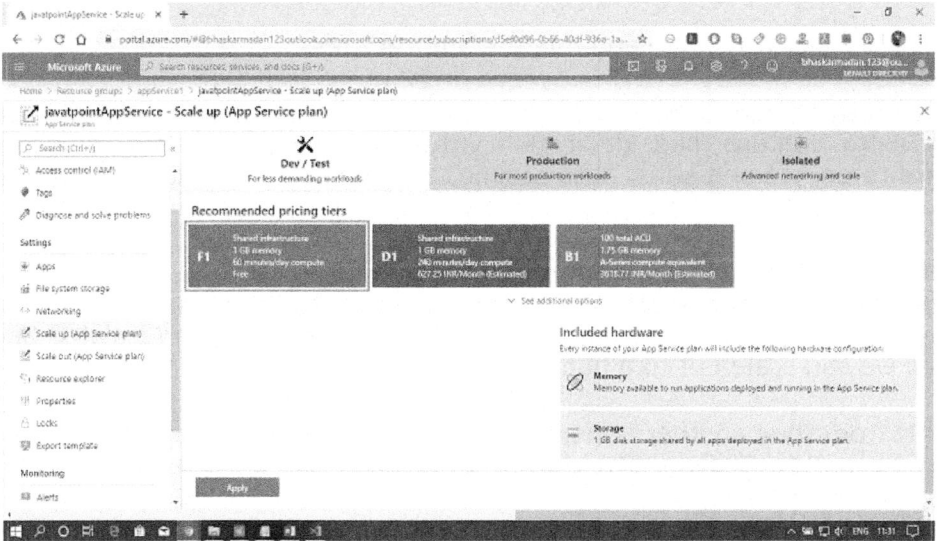

Step 3: Select the pricing tier then click on apply.

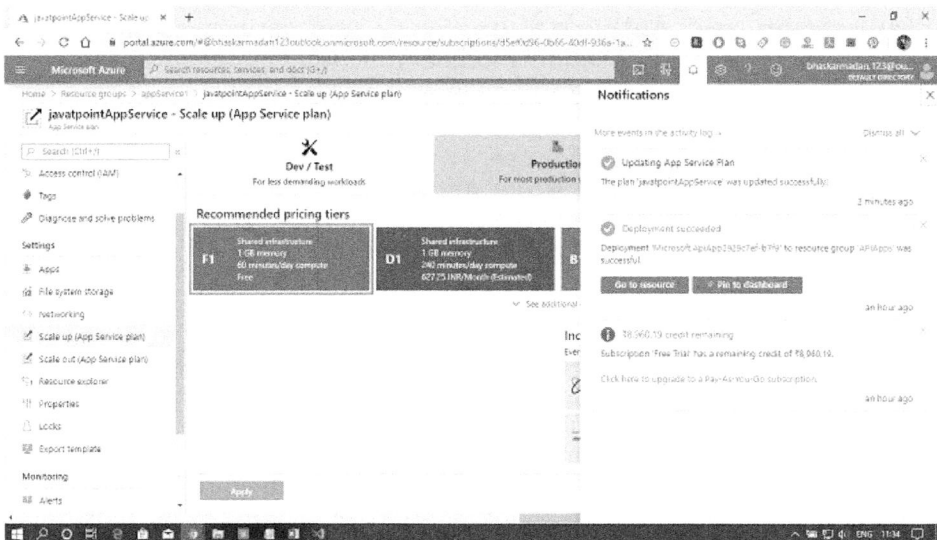

Step 4: It is successfully deployed, and similarly, you can scale out and set up it for auto-scaling based on some predefined conditions.

Azure App Service Security

Authentication and Authorization: Every App service comes with an Authentication and Authorization module that handles several things for our app.

* ❖ It will authenticate the user with a specified provider such as Facebook, Google, Twitter, Azure Active Directory, etc.
* ❖ It will store, validate, and refreshes tokens.
* ❖ It also manages the authenticated session.
* ❖ It injects identity formation into request headers.

How Authentication and Authorization works

First, the request from the client browser will come to the App service front-end. From that, the request will be forwarded to the Authentication and Authorization module. And that Authentication module will include all the Authorization and Authentication logic, which includes token management and also session management, etc., and it sits outside the web app code. That is the reason we don't need to change code between our web application to enable Authentication and Authorization for our app in Azure. We can able to slightly influence this Authentication and Authorization logic using the environment variable in terms of tracing.

Authentication and Authorization module handles several things for our app:

* ❖ Authenticates users with the specified provider
* ❖ Validate, store, and refreshes tokens
* ❖ Manages the authenticated session
* ❖ Injects identity information into request headers
* ❖ Logging & tracing

Other security areas

There are additional security areas that we need to be aware of for App service, which we can take advantage of them.

* ❖ **ISO, SOC, and PCI complaint:** If we are processing credit card information, the underlying environment is PCI compliant, but at the same time, you have to go for PCI compliance from the application layer perspective. But, from the environment perspective, Microsoft Azure App Service is PCI compliant
* ❖ **IP Address whitelisting:** In case if we want to limit the trigger to our App services form a specific trusted IP Address, then we can white list the same within the Azure portal for our app services.

❖ **SSL communication:** To encrypt the data at transit, we can enable SSL communication.

Managed Service Identity

This service is recently added to Azure. What we are going to do here is creating an identity for our app, and providing access to different services to that identity. By doing this, we don't need to store any userID-password to access certain Azure services. What we generally do is we go to the Azure portal and tell to ARM to create managed service identity for your Azure App Service. And when we trigger that, a service principle gets created in Azure active directory.

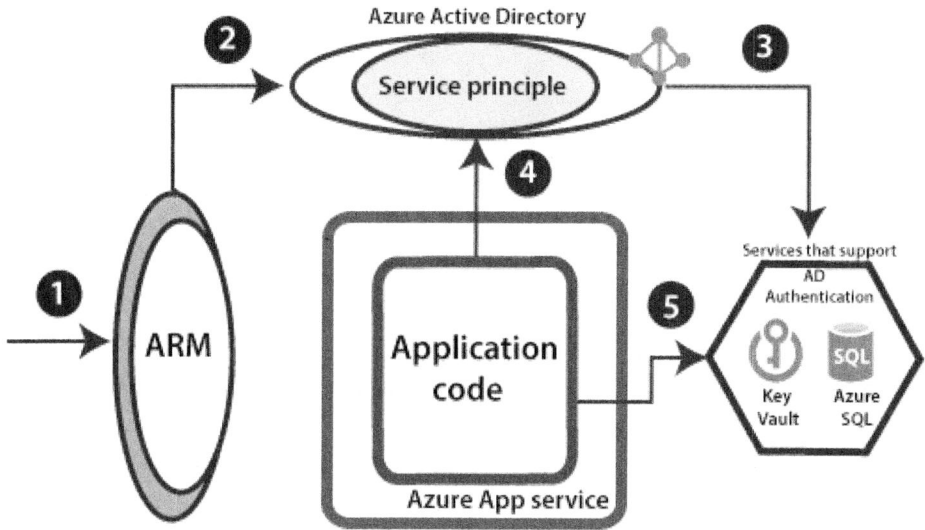

Example - if we want to Access a secret from Azure key vault. By submitting that token and having a proper access policy defined within Azure key vault, our application code will be able to retrieve the secret at run time and use that secret to access an on-premises resource.

App Service Environments security

If we are using App service environments, then we will get additional benefits in terms of security.

❖ **Network security groups:** We can associate with network security groups and control the traffic coming into our App service using network security groups.
❖ **Web Application Firewall:** It is a feature of application Gateway that provides centralized protection of your web applications from

common exploits and vulnerabilities. The web application firewall is based on rules from the OWASP core rule sets 3.0 or 2.2.9

Enabling authentication with Azure active directory for Web App

Step 1: Open your API App and click on Authentication/Authorization.

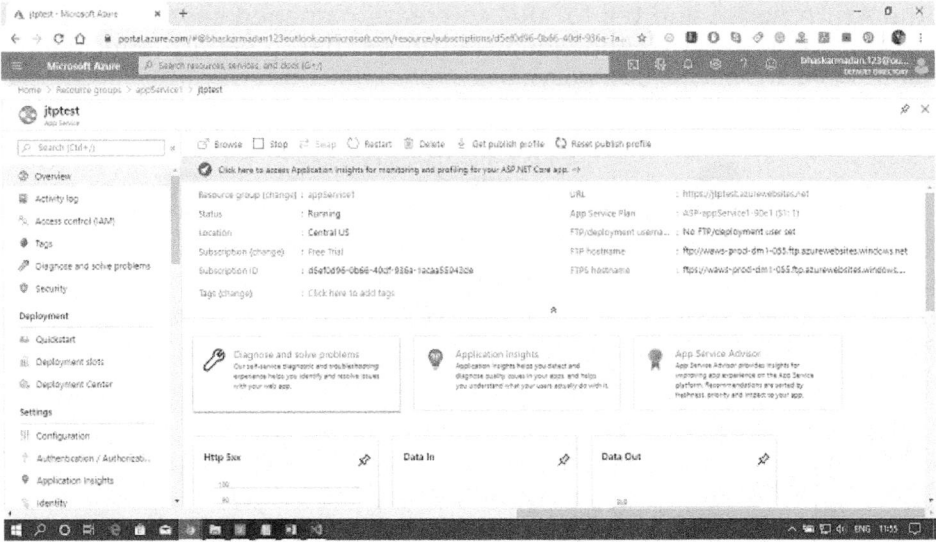

Step 2: Click on the toggle button showing switch on/off. Switch it on.

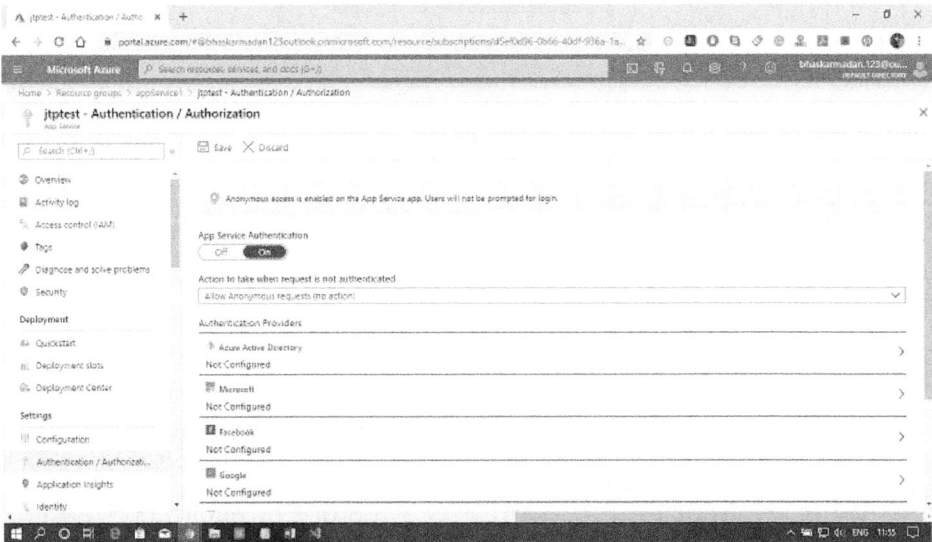

Step 3: Now select the Action to take when the request is not authenticated as "Login with Azure Active Directory."

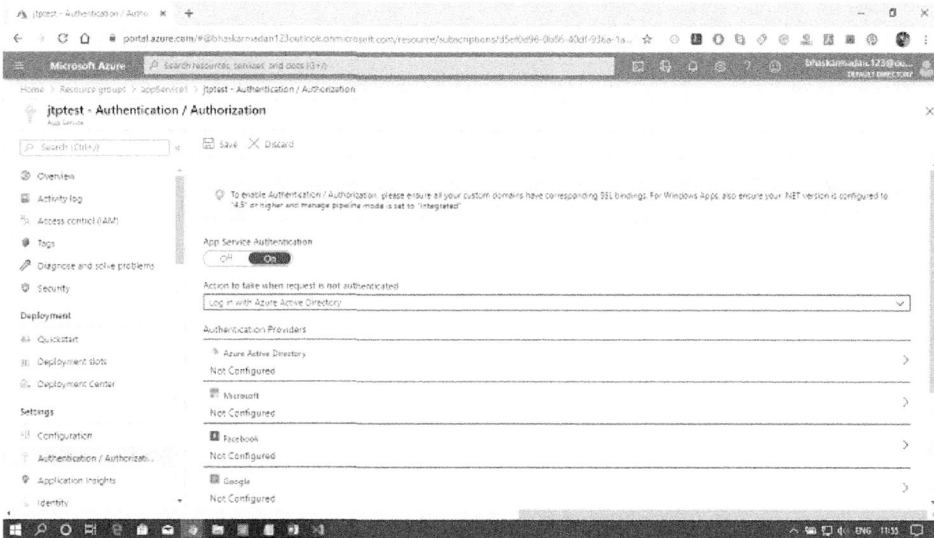

Step 4: Now, configure the Azure Active Directory with the express mode. After that, click on create and then click on save.

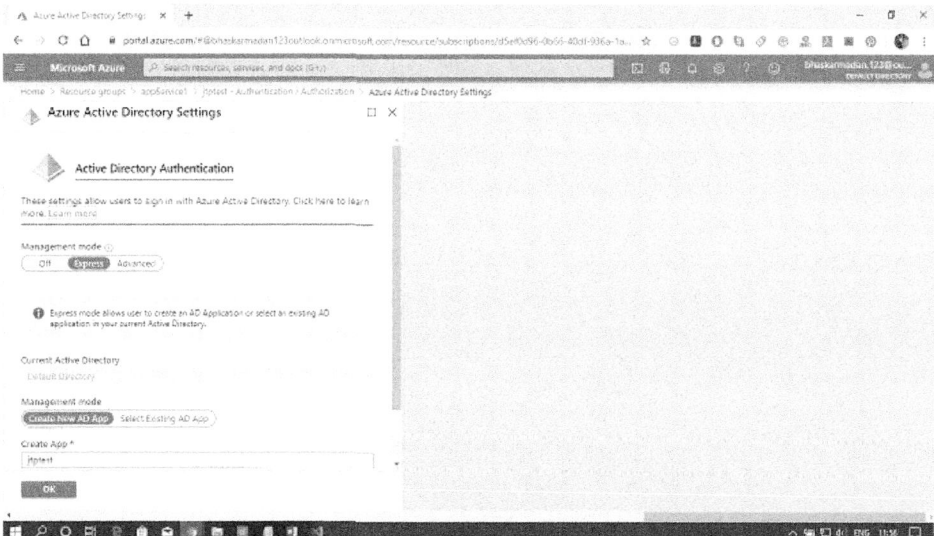

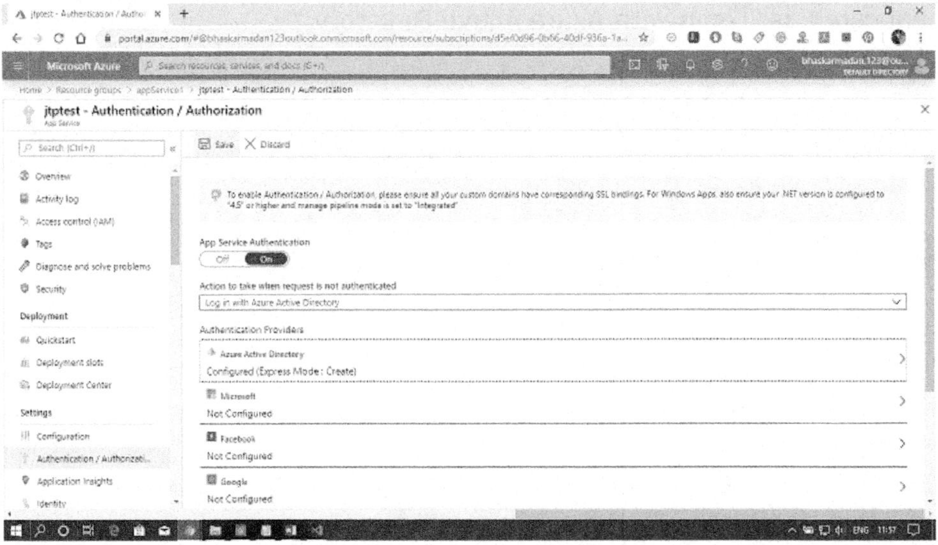

Azure App Service Monitoring

Microsoft provides different features to monitor our App service.

Quotas

Application hosted in App Service is subject to certain limits on the resources they can use. The boundaries are defined by the App Service plan associated with the app. When we create an app service plan, we generally select a pricing tier and also size. Based on the size and pricing tier, we'll be allocated with a certain amount of CPU and a certain amount of memory. Quotas for Free or Shared apps are as follows.

- ❖ CPU(Short)
- ❖ CPU(Day)
- ❖ Memory
- ❖ Bandwidth
- ❖ Filesystem- Applicable for Basic, Standard, and Premium plans

The only quota applicable to apps hosted on Basic, Standard, and Premium plans is Filesystem, based on the number of App service plans that we have created in our subscription.

Metrics

Metrics provide information about the app or App Service plan's behaviour. So Metrics are provided at two levels.

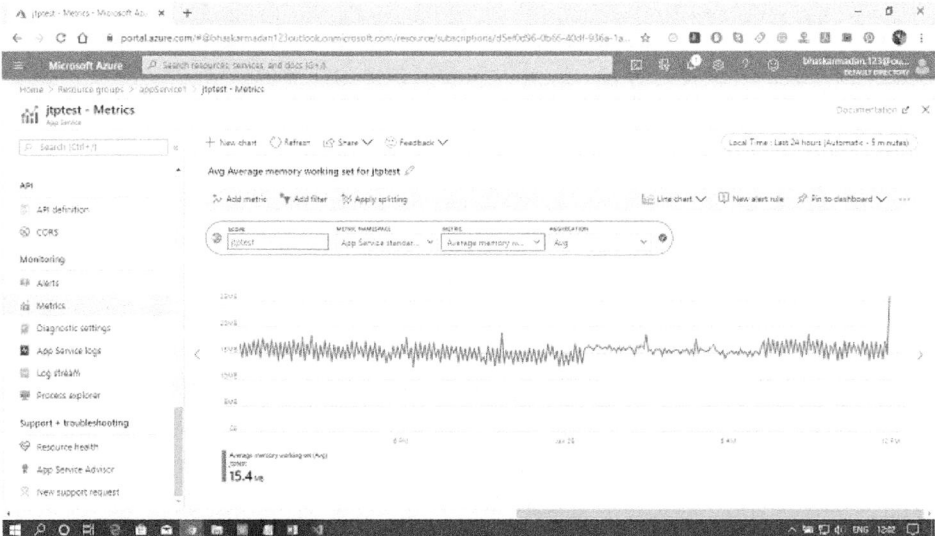

❖ One is at the App service plan level, which will include CPU and Memory percentage, Data in and out, Disc queue length, and HTTP queue length.

❖ The second level of metrics provided at an application level, which includes average response time, average memory working set, CPU time, Data in and out, etc. All of these metrics are essential. There are a lot more metrics available when we go for application insight, which is an advanced monitoring tool using which we can monitor web apps.

Granularity and retention

❖ Minute granularity: These metrics are retained for 30 hours.
❖ Hour granularity: These metrics are retained for 30 days.
❖ Day granularity: These metrics are retained for 30 days.

However, if we use the application insights, then this retention will increase drastically.

Diagnostics

App service web apps provide diagnostic functionality for logging information from both the web server and the web application. These are logically separated into web server diagnostics and application diagnostics.

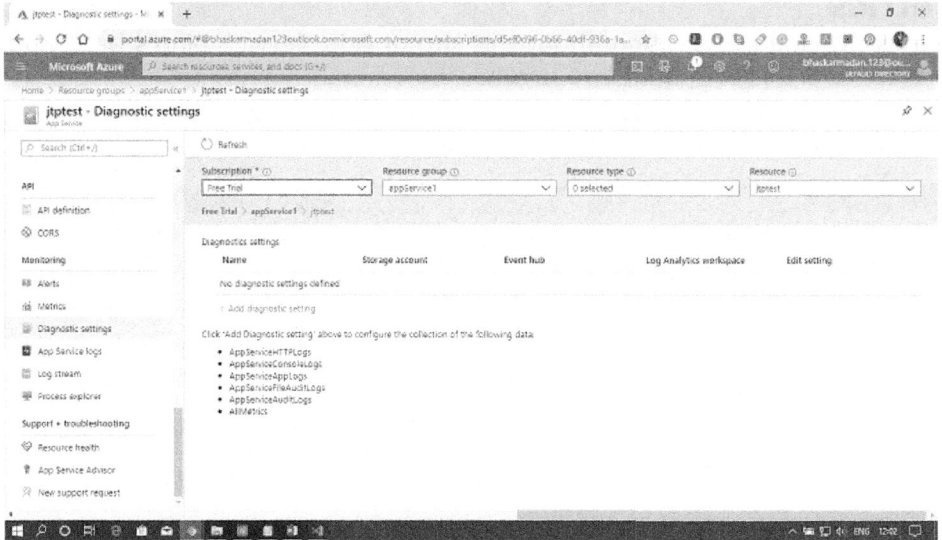

Web server Diagnostics: In this, we can have a detailed error logging, which is any Http 400, and the above error will get logged. We can also view the logs. And the second thing is failed request tracking, which basically contains the trace of IIS components that are used to process our request. And the third one is web server logging. These are the overall metrics on how many Http requests we received, and from a particular IP address.

Application Diagnostics: It allows us to capture the information produced by the web application. So if we're using system diagnostic trace, all the information will be provided by application diagnostics.

Diagnostic information can be stored in file system or Azure storage.

Stream logs

During the development of an application, it is often useful to see logging information in near-real-time. We can steam logging information to our development environment using either Azure PowerShell or the azure Command-line interface.

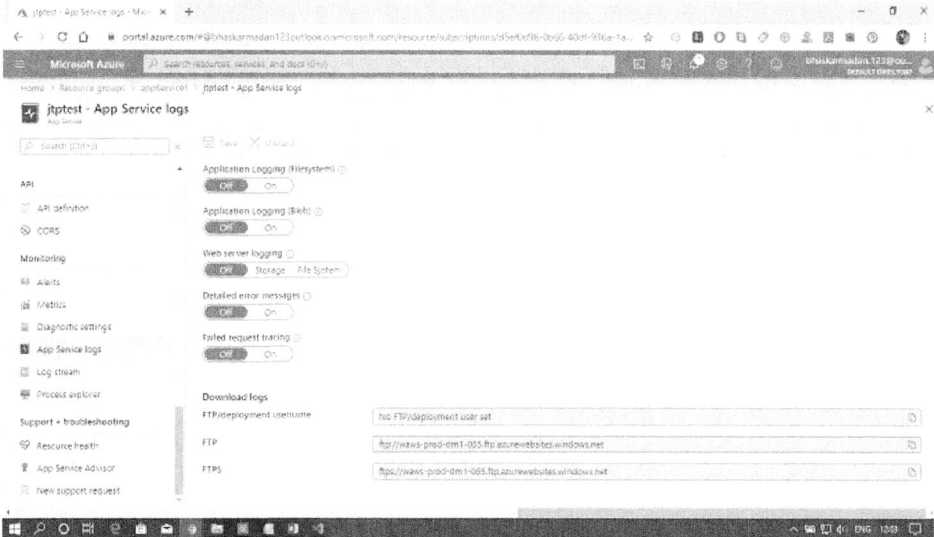

These are the basic commands that we need to use to stream the log files.

❖ Streaming with Azure PowerShell

    ❖ Get-AzureWebSiteLog - Name webappname -Tail
    ❖ Get-AzureWebSitelog - Name webappname -Tail -Message Error

❖ Streaming with Azure Command-Line Interface

- ❖ az webapp log tail - name webappname -resource-group myResourceGroup
- ❖ az webapp log tail -name webappname -resource-group myResourceGroup -filter Error

Azure Content Delivery Network

Azure CDN caches web content at a strategically placed location to provide maximum throughput for delivering content to users. To better explain this, let?s take an example.

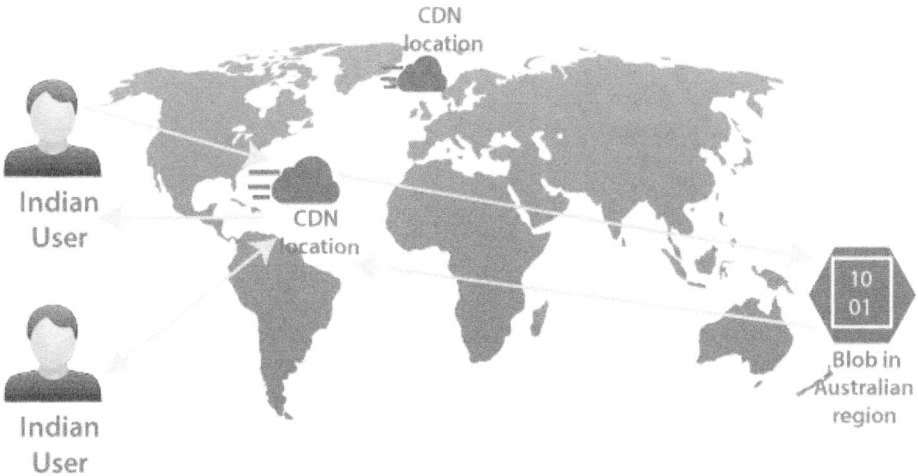

Let?s say we have a vast amount of video content located in Australia, but the primary users of that content are located in the US, and if any of the users from India will try to access the content from Australia. Then they will experience some latency because of the distance between Australia and India. In that scenario, we can use a content delivery network to reduce that latency.

CDN products
There are several types of products that are available by Azure, and there are two other third party providers that provide CDN products in partnership with Microsoft.

- ❖ Azure CDN Standard from Microsoft (Preview)
- ❖ Azure CDN Standard from Akamai
- ❖ Azure CDN Standard from Verizon
- ❖ Azure CDN Premium from Verizon

Features of Content Delivery Network (CDN)
Following are the fundamental features of Azure CDN:

- ❖ Dynamic site acceleration: It is the capability to deliver dynamic web content with minimum latency. It is achieved by using different techniques such as route optimization to avoid congestion points, TCP optimization, etc.

❖ HTTPS support: It provides us the HTTPS support of secure web content.

❖ Query string caching: Based on query string caching, we can cache the content also within CDN location.

❖ Geo-Filtering: We can apply some geo-filtering if we want certain content filtered for a particular geographical region.

❖ Azure diagnostics logs: It provides the facility of records of diagnosis.

CDN configuration

❖ When we start using CDN, the first thing we will create is the CDN profile. It is a collection of CDN endpoints, and by default, it can contain up to 10 CDN endpoints. When we are creating a CDN profile, we will specify the type of product that you want to use. For Example, CDN premium from Verizon or CDN standard for Microsoft, etc.

❖ Secondly, we will create a CDN endpoint. When we are creating CDN endpoint, we will specify the name, and also origin type what exactly we?re trying to configure this CDN for. It can be Azure storage, cloud storage, web app, or a custom origin.

❖ Finally, we will define the Origin path where these videos or web content is located and also the protocol of origin. Once we create a CDN endpoint, we?ll get an endpoint that will be whatever the name we have given ?example.net.?

Ways to control how files are cached

❖ Caching rules

    ❖ Global caching rules
    ❖ Custom caching rules

❖ Purged cached assets
❖ Pre-load assets on an Azure CDN endpoint

Azure Media Service

It is an extensible cloud-based platform that enables developers to build scalable media management and delivery applications. For example - if we want to develop an app like DailyMotion, then we can do so by using Microsoft Azure media services.

Azure media services are based on REST APIs that enable us to securely upload, store, encode, and wrap video or audio content for both on-demand and live stream delivery to various clients. Those clients can be TV, PC, and mobile devices also.

Media Services Concepts
* **Assets:** An Asset contains digital files and the metadata about these files. These files can be audio, video or image, etc.
* **AssetFile:** It contains metadata about the media file.
* **AccessPolicy:** It defines the permission and duration of access to an asset.
* **Locators:** It provides an entry point to access the files contained in an asset.
* **Job:** It is used to process one audio/video presentation.
* **Channels:** It is responsible for processing live streaming content. It provides an input endpoint that is provided to a live transcoder.
* **Program:** It enables us to control the publishing and storage of segments in a live stream.
* **Streaming endpoint:** It represents a streaming service that delivers content.

The architecture of Media Service
* **Delivering on-demand:** In this case, first, we will upload a high-quality media file into an asset, and then we encode it to a set of adaptive bit that reads MP4 files. After that, we configure the asset delivery policy. Asset delivery policy tells Media services how we want our assets to be delivered using which protocol. Now, we will publish an asset by creating an on-demand locator and stream the published content.

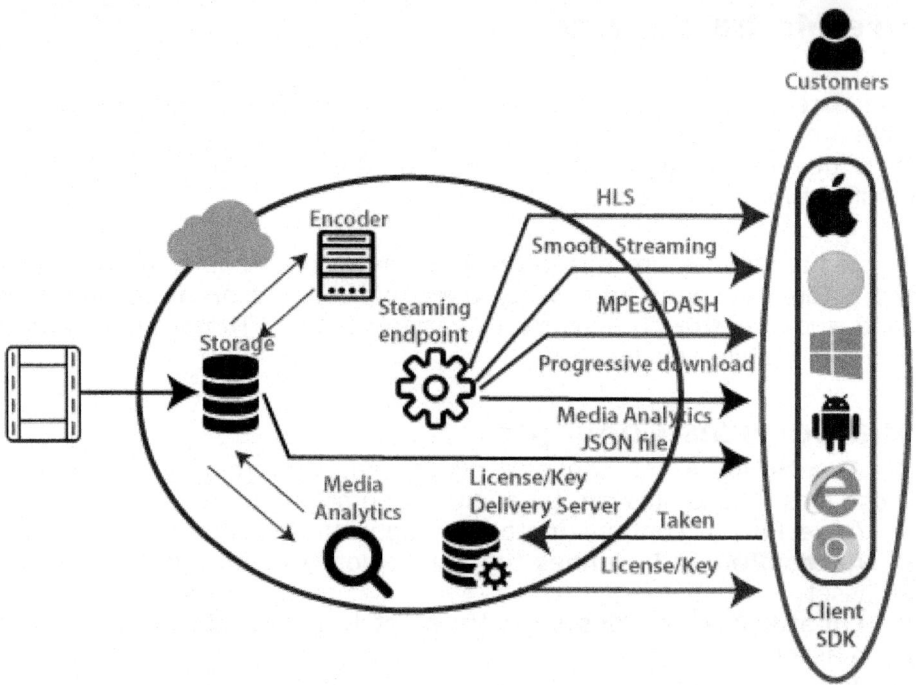

❖ **Live-Streaming:** We can broadcast live content using various live streaming protocols. We might go to encode our stream into an adoptive bit read stream. We can preview our live stream also. Finally, we can deliver the content through common streaming protocols such as Smooth, HLS, etc.

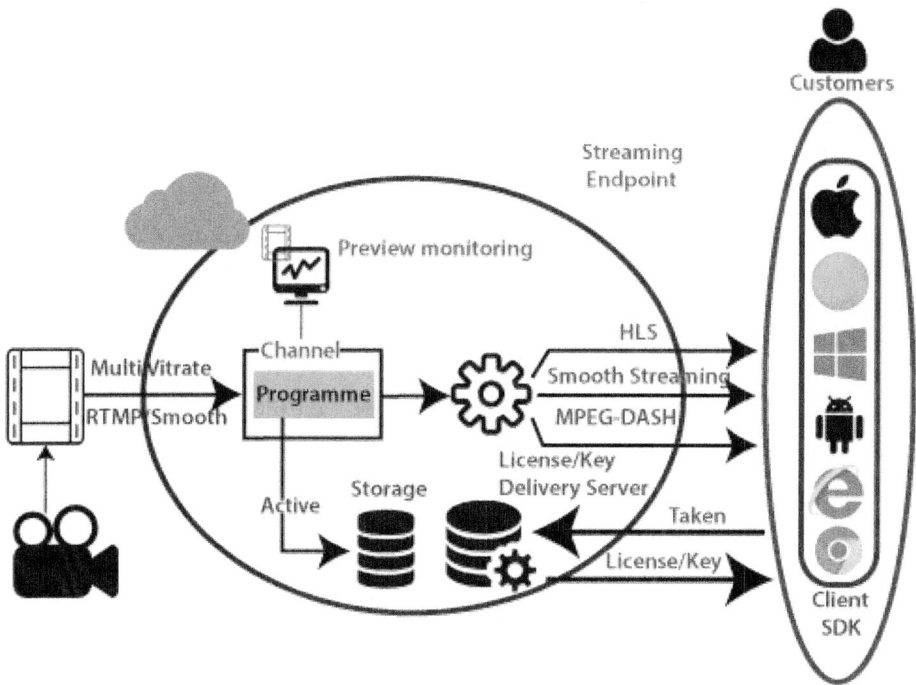

Azure Search

Azure Search is a cloud Search as a Service that enables us to add a robust search experience to our applications using a simple REST API or .NET SDK, without managing a search infrastructure.

Features of Azure Search

- ❖ Powerful queries
- ❖ Multi-language support
- ❖ Search suggestions
- ❖ Hit highlighting
- ❖ Faceted Navigation

Above are the different features associated with Azure search. In case if we want to have a cloud-based search engine that we can embed in our web application. Azure offers a service called Azure Search.

Section VI: Azure Database service

The basic fundamental building block that is available in Azure is the SQL database. Microsoft offers this SQL server and SQL database on Azure in many ways. We can deploy a single database, or we can deploy multiple databases as part of a shared elastic pool.

Azure Database Service Architecture

Microsoft introduced a managed instance that is targeted towards on-premises customers. So, if we have some SQL databases within our on-premises datacentre and we want to migrate that database into Azure without any complex configuration, or ambiguity, then we can use managed instance. Because this is mainly targeted towards on-premises customers who want to lift and share their on-premises database into Azure with the least effort and optimized cost. We can also take advantage of licensing we have within our on-premises data center.

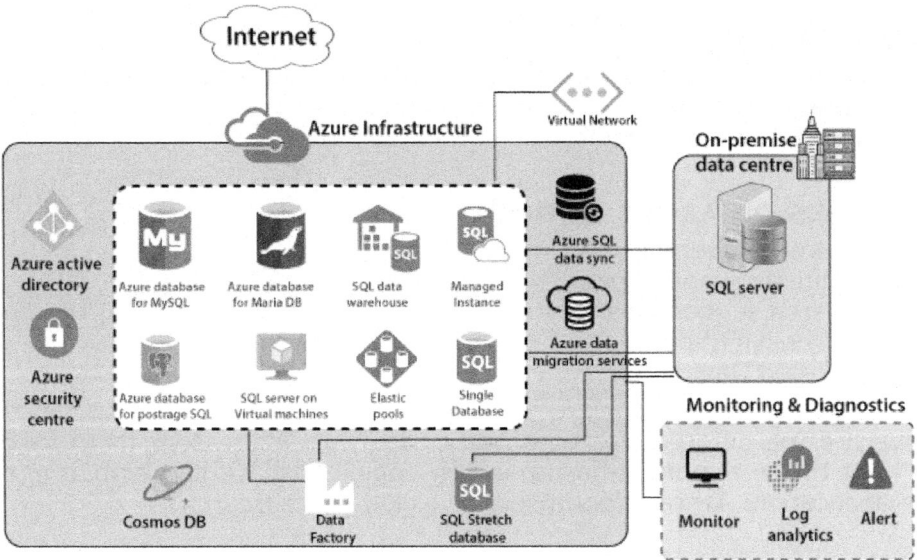

Microsoft will be responsible for maintenance patching and related services. But, in case if we want to go for the IaaS service for the SQL server, then we can deploy SQL Server on the Azure Virtual machine. If the data have a dependency on the underlying platform and we want to log into the SQL Server, in that case, we can use the SQL server on a virtual machine.

We can deploy a SQL data warehouse on the cloud. Azure offers many other database services for different types of databases such as MySQL, Maria DB, and also PostgreSQL. Once we deployed a database into Azure, we need to migrate the data into it or replicate the data into it.

Azure Database Services for Data Migration

The services that are available in Azure, which we can use to migrate the data from our on-premises SQL Server into Azure.

Azure Data Migration Service: It is used to migrate the data from our existing SQL server and database within the on-premises data center into Azure.

Azure SQL data sync: If we want to replicate the data from our on-premises database into Azure, then we can use Azure SQL data sync.

SQL Stretch Database: It is used to migrate cold data into Azure. SQL stretch database is a bit different from other database offerings. It works as a hybrid database because it divides the data into different types - hot and cold. A hot data will be kept in the on-premises data center and cold data in the Azure.

Data Factory

It is used for ETL transformation, extraction loading, etc. Using the data factory, we can even extract the data from our on-premises data center. We can do some conversion and load it into the Azure SQL database. Data Factory is an ETL tool that is offered on the cloud, which we can use to connect to different databases, extract the data, transform it, and load into a destination.

Security

All the databases that are existing in Azure need to be secured, and also we need to accept connections from known origins. For this purpose, all these database services come with firewall rules where we can configure from which particular IP address we want to allow connections. We can define those firewall rules to limit the number of connections and also reduce the surface attack area.

Cosmos DB

Cosmos DB is a NoSQL data store that is available in Azure, and it is designed to be globally scalable and also very highly available with extremely low latency. Microsoft guarantees latency in terms of reading and writes with Cosmos DB. For example - if we have any applications such as IoT, gaming where we get a lot of data from different users spread across globally, then we will go for Cosmos DB. Because Cosmos DB is designed to be globally scalable and highly available due to which our users will experience low latency.

Finally, there are two things, and one is we need to secure all the services. For that purpose, we can integrate all these services with Azure Active Directory and manage the users from Azure Active Directory also. To monitor all these services, we can use the security center. There is an individual monitoring tool too, but Azure security center will keep on monitoring all these services and provide recommendations if something is wrong.

Azure SQL Database

SQL database is the flagship product of Microsoft in the database area. It is a general-purpose relational database that supports structures like relation data - JSON, spatial, and XML. The Azure platform fully manages every Azure SQL Database and guarantees no data loss and a high percentage of data availability. Azure automatically handles patching, backups, replication, failure detection, underlying potential hardware, software or network failure, deploying bug fixes, failovers, database upgrades, and other maintenance tasks.

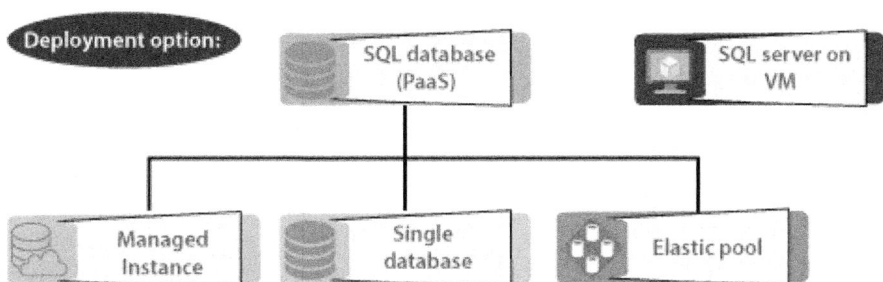

There are three ways we can implement our SQL database

❖ **Managed Instance:** This is primarily targeted towards on-premises customers. In case, if we already have a SQL server instance in our on-premises data-center and you want to migrate that into Azure with minimum changes to our application and the maximum compatibility. Then new will go for the managed instance.
❖ **Single database:** We can deploy a single database on Azure its own set of resources managed via a logical server.
❖ **Elastic pool:** We can deploy a pool of databases with a shared set of resources managed via a logical server.

We can deploy the SQL database as an infrastructure as a service. That means we want to use the SQL server on an Azure virtual machine, but in that case, we are responsible for managing the SQL server on that particular Azure virtual machine.

Purchasing model
There are two ways we can purchase the SQL Server on Azure.

❖ **VCore purchasing model:** The vCore-based purchasing model enables us to independently scale compute and storage resources, match on-premises performance, and optimize price. It also allows us to choose a generation of hardware. It also allows us to use Azure Hybrid Benefit for SQL Server to gain cost savings. Best for the customer who values flexibility, control, and transparency.

❖ **DTU model:** It is based on a bundled measure on compute, storage, and IO resources. Sizes of the compute are expressed in terms of Database Transaction Units (DTUs) for single databases and elastic Database Transaction Units (eDTUs) for elastic pools. This model is best for customers who want simple, pre-configured resource options.

Azure SQL Database service tiers

❖ **General Purpose/ Standard model:** It is based on a separation of computing and storage service. This architectural model depends on the high availability and reliability of Azure Premium Storage that transparently copies database files and guarantees for zero data loss if underlying infrastructure failure happens.

❖ **Business Critical/ Premium service tier model:** It is based on a cluster of database engine processes. Both the SQL database engine process and underlying mdf/ldf files are placed on the same node with locally attached SSD storage providing low latency to our workload. High availability is implemented using technology similar to SQL Server Always On Availability Groups.

❖ **Hyperscale service tier model:** It is the newest service tier in the vCore-based purchasing model. This tier is a highly scalable storage and computes performance tier that leverages the Azure architecture to scale-out the storage and computes resources for an Azure SQL Database beyond the limits available for the General Purpose and Business Critical service tiers.

SQL database logical server

❖ It acts as a central administrative point for multiple single or pooled database logins, firewall rules, auditing rules, threat detection policies, and failover groups.

❖ It must exist before we can create the Azure SQL database. All databases on a server are created within the same region as the logical server.

❖ The SQL database service makes no guarantees regarding the location of the database in relation to their logical servers and exposes no instance-level access or features.

❖ An Azure database logical server is the parent resource for databases, elastic pools, and data warehouses.

Elastic pools

❖ It is a simple and cost-effective solution for scaling and managing more than one database. The databases inside an elastic pool are on a single Azure SQL Database server and share a group of resources at a fixed price.

❖ We can configure resources for the pool based either on the DTU-based purchasing model or the vCore-based purchasing model.

❖ The size of a pool always depends on the aggregate resource needed for all databases in the pool. It determines the following options:

> ❖ The maximum resources utilized in the pool by the databases.
> ❖ The maximum storage bytes utilized in the pool by the databases.

Creating an Azure SQL Database using Azure portal

Step 1: Click on *create a resource* and search for SQL Database. Then click on create.

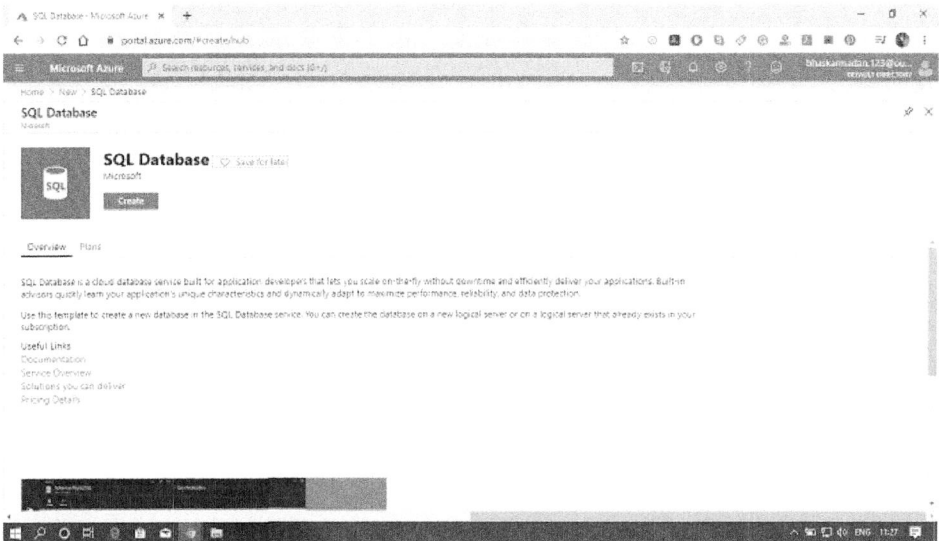

Step 2: Fill all the required details.

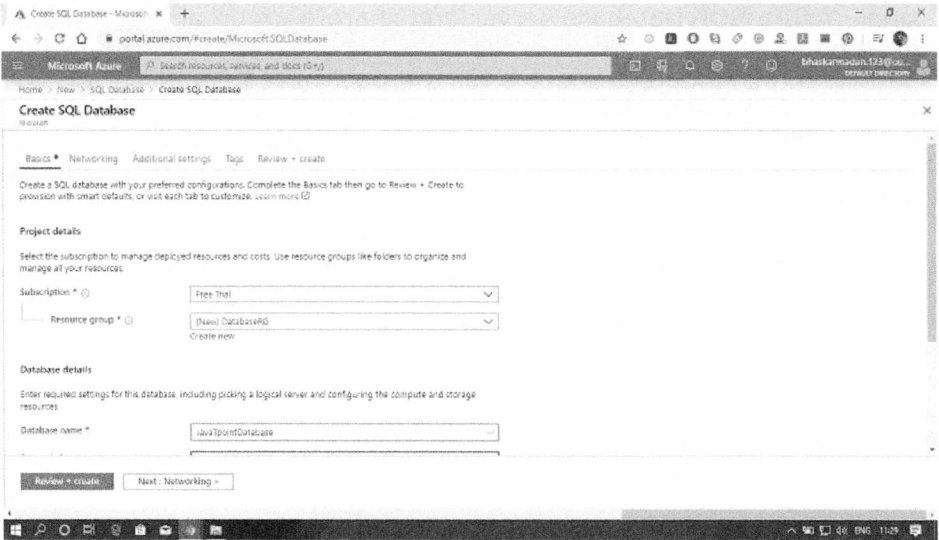

Step 3: Select a server or create a new one, as shown in the figure given below.

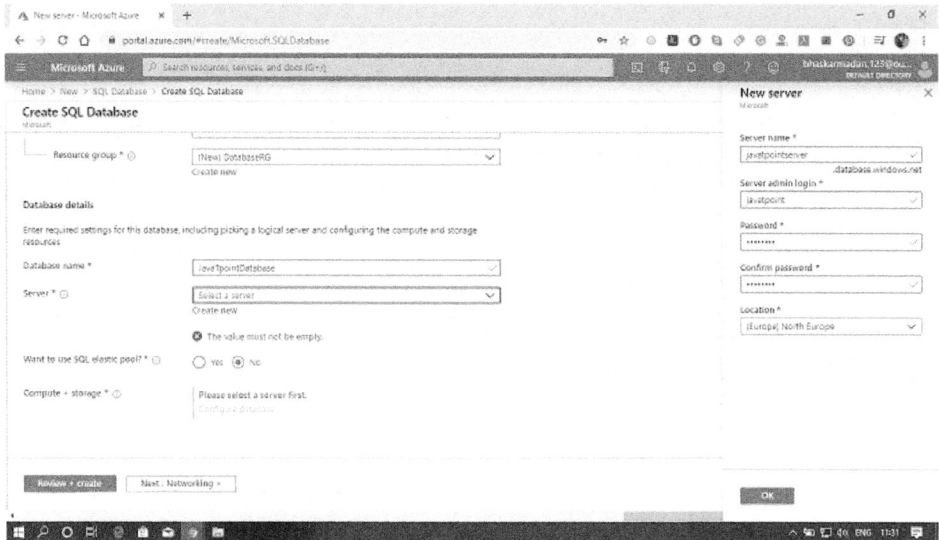

Step 4: Now, select the pricing tier by clicking on Compute + Storage, as shown in the figure below.

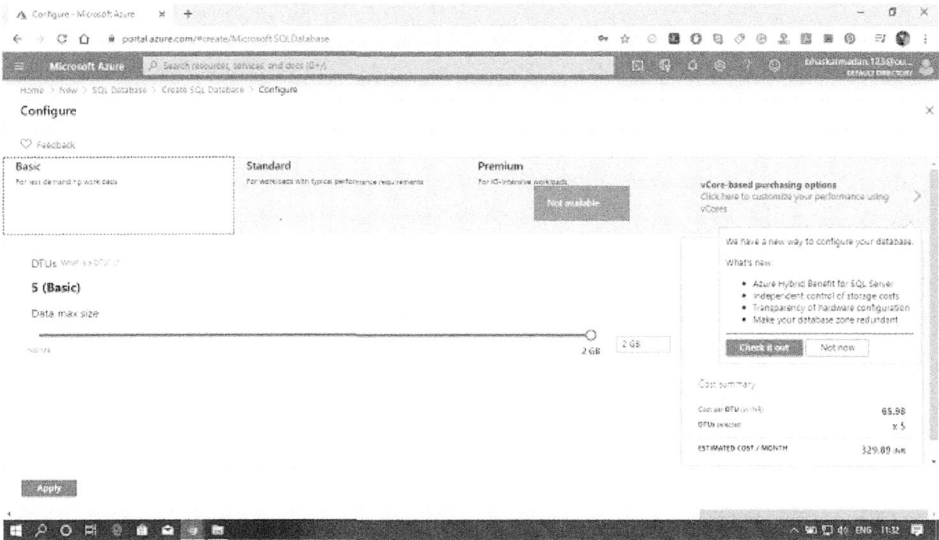

Step 5: After that, click on Review + Create and create the SQL database for your apps.

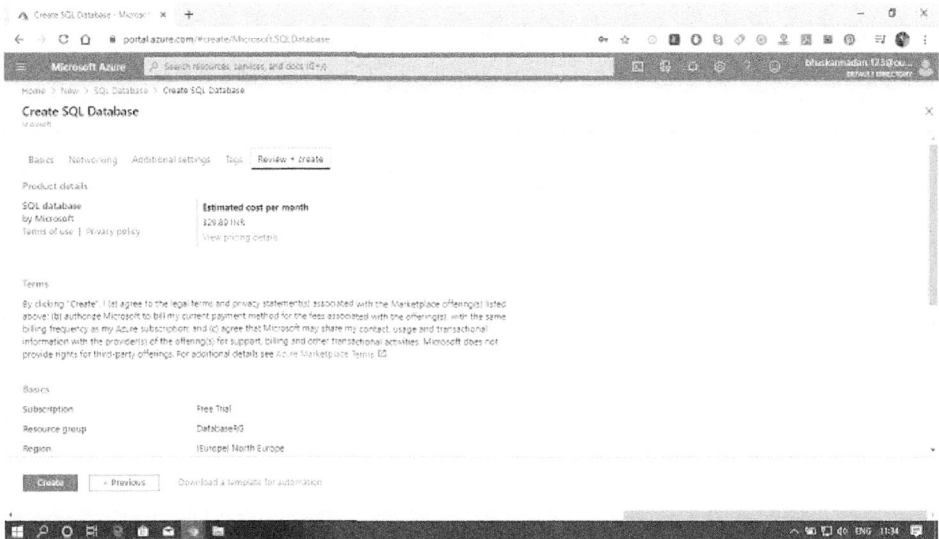

Step 6: Your SQL database is now created, now click on the go-to resources to configure additional settings for your database.

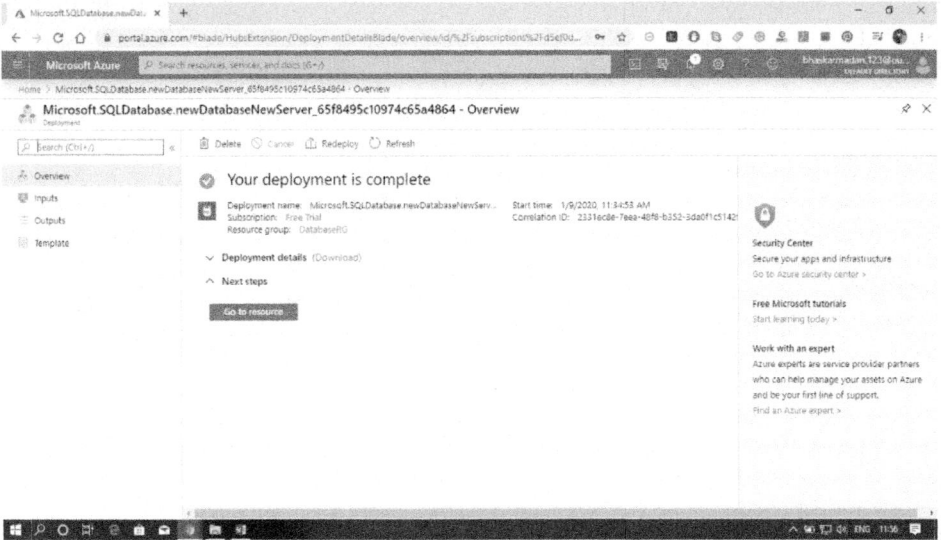

Azure SQL Database Configuration

We'll see here the key configuration features of the Azure SQL Server and SQL database. In terms of Azure SQL database configuration, the first key thing is Firewall rules at a server level.

Firewall Rules

At a logical server within Azure, we can define some firewall rules. It can be IP rules. IP rules will grant access to the database based on the originating IP address of each request. And the second type of rule is the virtual network rule. It is based on virtual network service endpoints.

Rules for Azure SQL databases can be defined at two levels:

❖ **Server level firewall rules:** These firewall rules enable clients to access our entire Azure SQL server, i.e., each database within a similar logical server. These firewall rules will be stored in the master database. Server-level firewall rules can be configured by using the portal or by using Transact-SQL statements.
❖ **Database-level firewall rules:** These rules enable clients to access certain (secure) databases within the same logical server. We can create these rules for each database (including the master database), and they are stored in the individual databases.

Configuring Firewall rules in Azure portal

Step 1: Go to the firewall setting in your database server that you have already created. After that, click on **Add Client**.

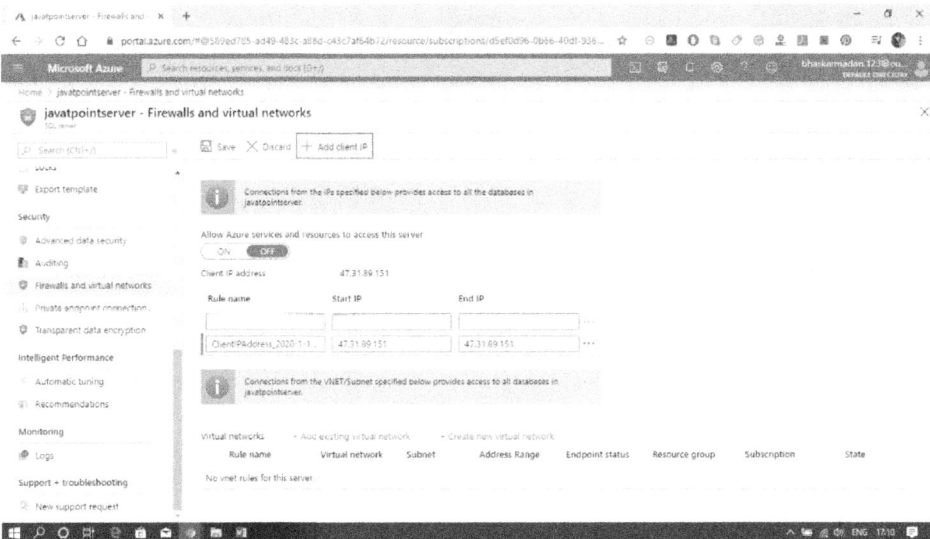

Step 2: Now, click on *Add existing virtual network* and fill the required details, as shown in the figure below.

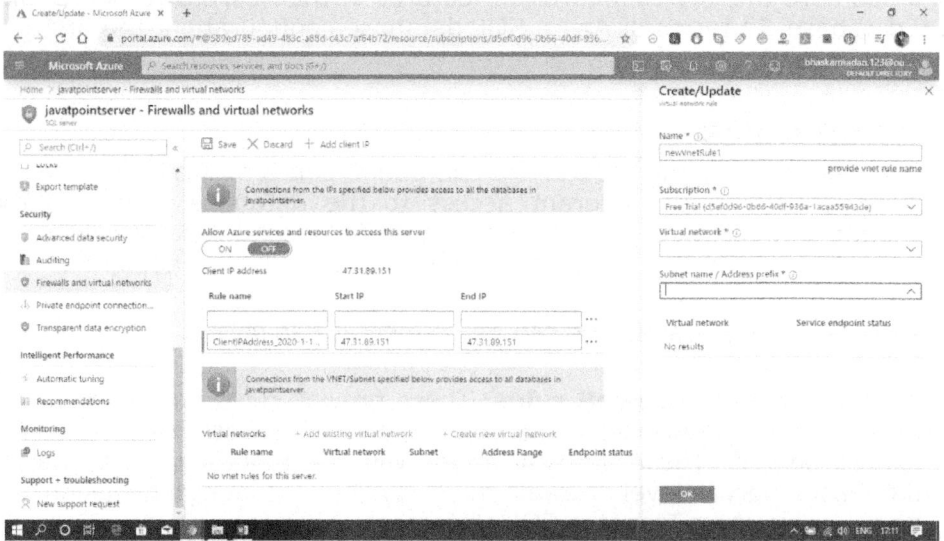

Step 3: Finally click on save, you will get the notification that your firewall rules got updated.

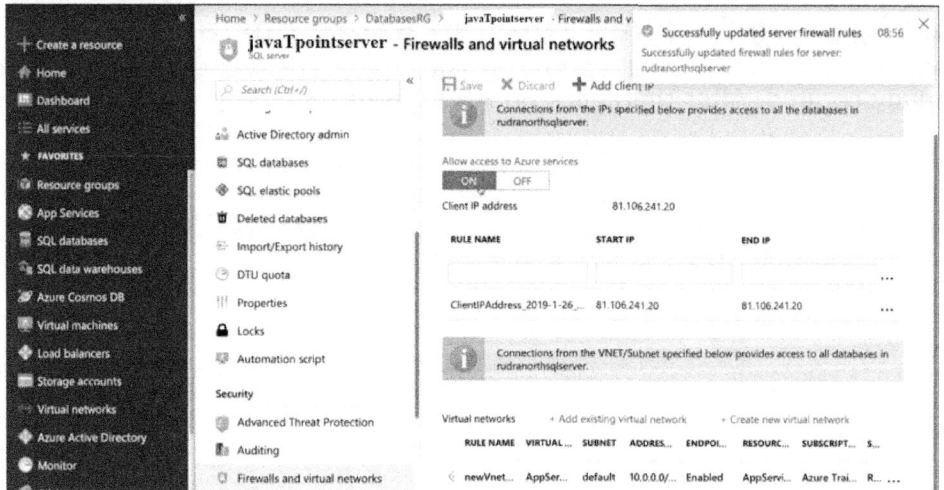

Geo-Replication

It is defined at a database level, not server level, and it is designed as a business continuity solution that allows the application to perform quick disaster recovery of individual databases in case of a regional disaster or large scale outage.

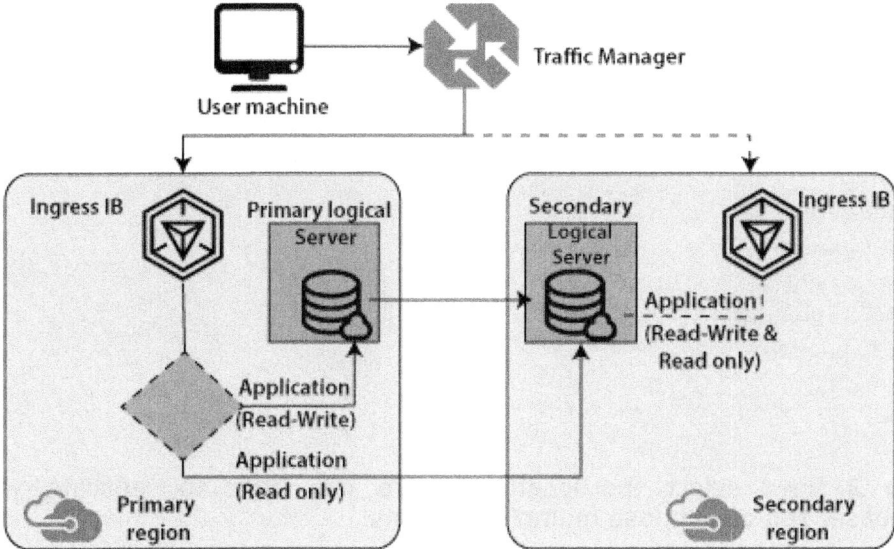

When we are configuring geo-replication, we specify a secondary database at a location far away from the primary location. We can have a traffic manager that routes the traffic by default to our primary load balancer and that the primary load balancer is based on the application request. If it is read and write, then it can route to a primary logical server. If it is 'read-only', it can route to a secondary server. Thereby the advantage of geo-replication is that we can offload some of the read-only traffic from primary and route to secondary.

The primary performance will be good because read-only queries will consume a certain amount of CPU or DTU units installed that we have a secondary database where the data continuously get replicated.

Configuring Geo-Replication using Azure portal
Step 1: Click on the Geo-Replication option; you will see the following window.

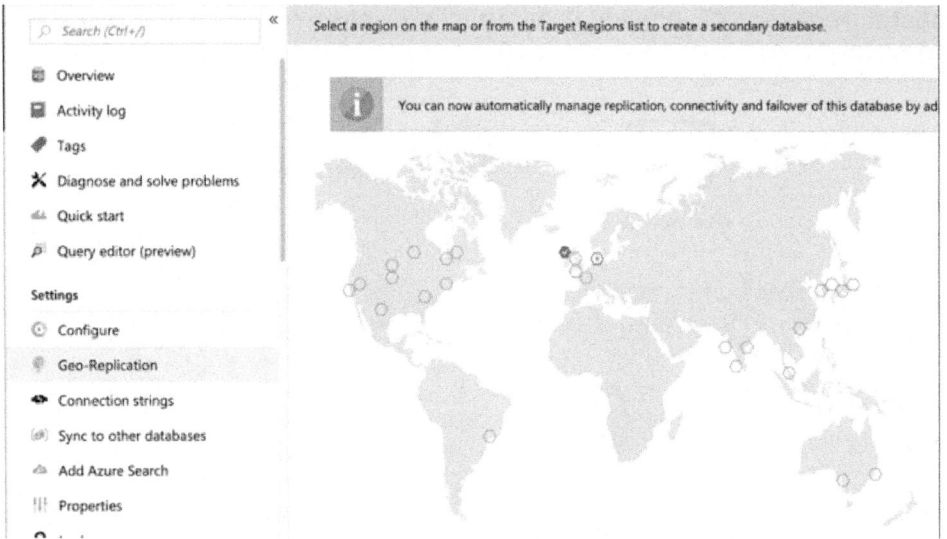

Step 2: Now, select the location where you want to replicate your database. You can choose multiple locations.

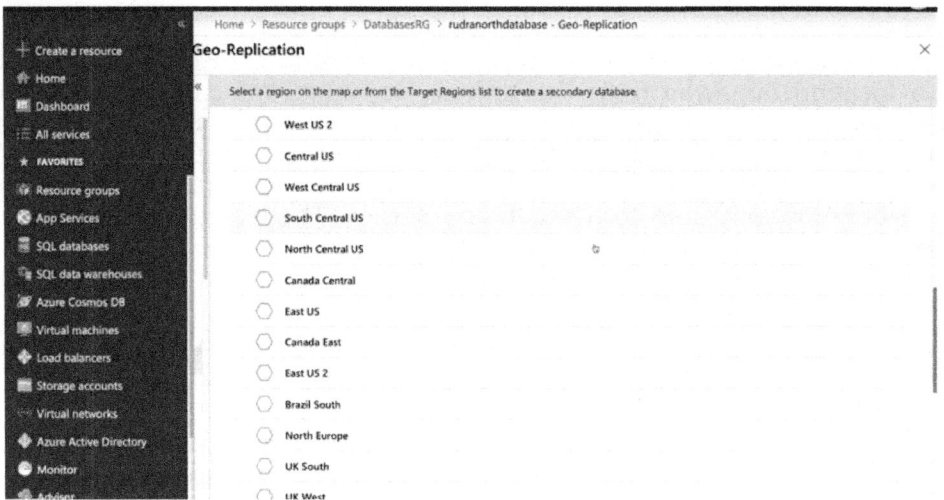

Step 3: Now, create a SQL server for the place where you want to replicate your data.

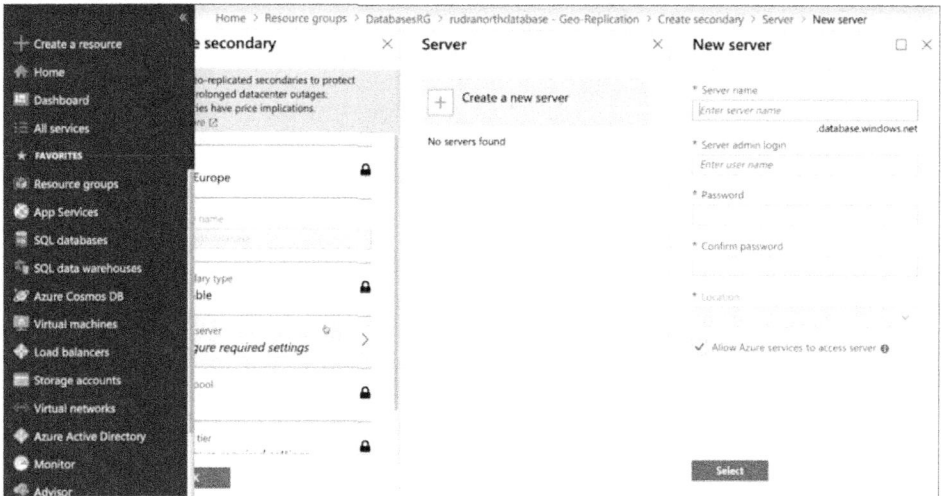

Step 4: Your server has been created and replicated successfully.

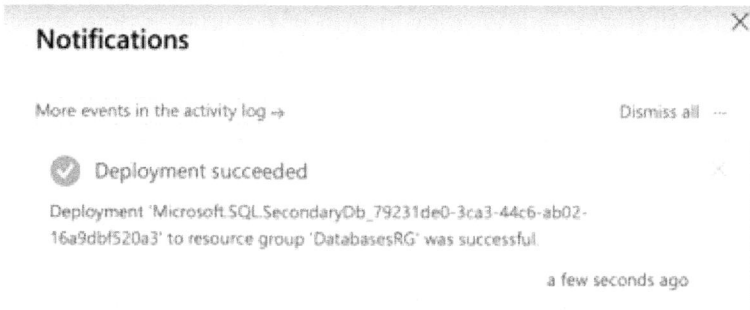

Step 5: You can see in the following figure, where the servers are replicated.

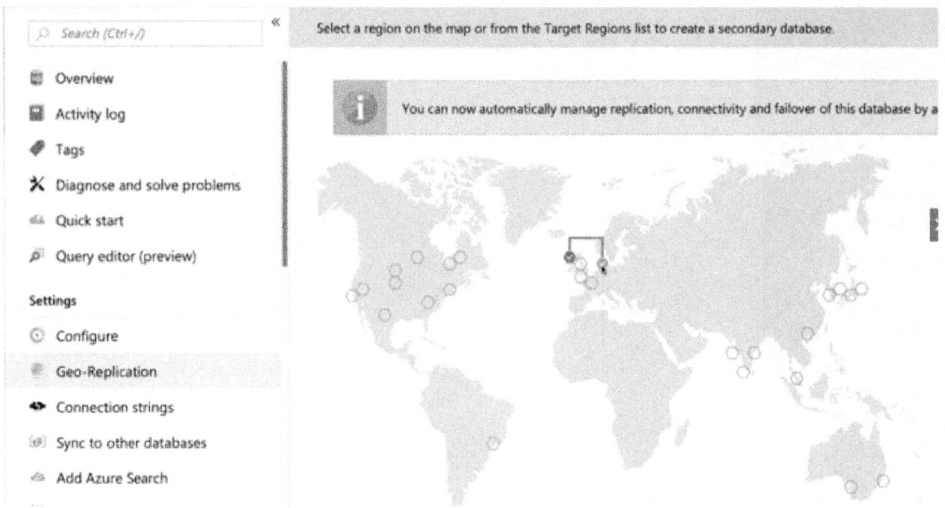

Failover Groups

Auto-failover group is a feature of the SQL database that allows us to manage replication and failover of a group of databases on a logical server or all databases in a Managed Instance to another region.

We can initiate failover manually, or we can delegate it to SQL Database service based on a user-defined policy. When we are using auto-failover groups with automatic failover policy, any outage that impacts one or many of the databases in the group results in automatic failover. It allows the read-write SQL application to transparently reconnect to the primary database when the database change after failover.

Database backups

The SQL database uses SQL server technology to create full, differential, and transaction log backups for Point-in-time Restore (PITR). The transaction log backups generally occur every 5-10 minutes, and differential backups occur typically every 12 hours, with the frequency based on the compute size and amount of database activity. Each SQL database has a default backup retention period between 7 and 35 days that depends on the purchasing model and service tier.

Long-term backup retention (LTR) leverages the full database backups that are automatically created to enable point-time restore. These backups are copied to different storage blobs if the LTR policy is configured. We can set an LTR policy for each SQL database and specify how frequently we need to copy the backups to the long-term storage blobs.

Azure SQL Managed Instance

The Azure SQL Database Managed Instance is a new implementation model of Azure SQL Database based on the VCore-based purchasing model.

Advantages of using Managed Instance

Easy lift and shift: Customers can lift and shift their on-premises SQL server to a Managed Instance that offers compatibility with SQL Server on-premises.

Fully managed PaaS: Azure SQL Database Managed Instance is designed for customers looking to migrate a large number of apps from on-premises self-built or ISV provided an environment to fully managed PaaS cloud environment.

New Business model: Competitive, transparent, and frictionless business model

Security: Managed Instance that offers compatibility with SQL Server on-premises and complete isolation of customer instances with native VNet support.

Managed Instance security isolation

Managed Instance provides additional security isolation from other tenants in the Azure cloud.

The managed instance security isolation includes:

❖ Native virtual network implementation and connectivity to our on-premises environment using Azure Express Route or VPN Gateway.
❖ SQL endpoint is exposed only through a private IP address, allowing safe connectivity from private Azure or hybrid networks.
❖ It is a Single-tenant environment with dedicated underlying infrastructure (compute, storage).

Structure of Managed Instance

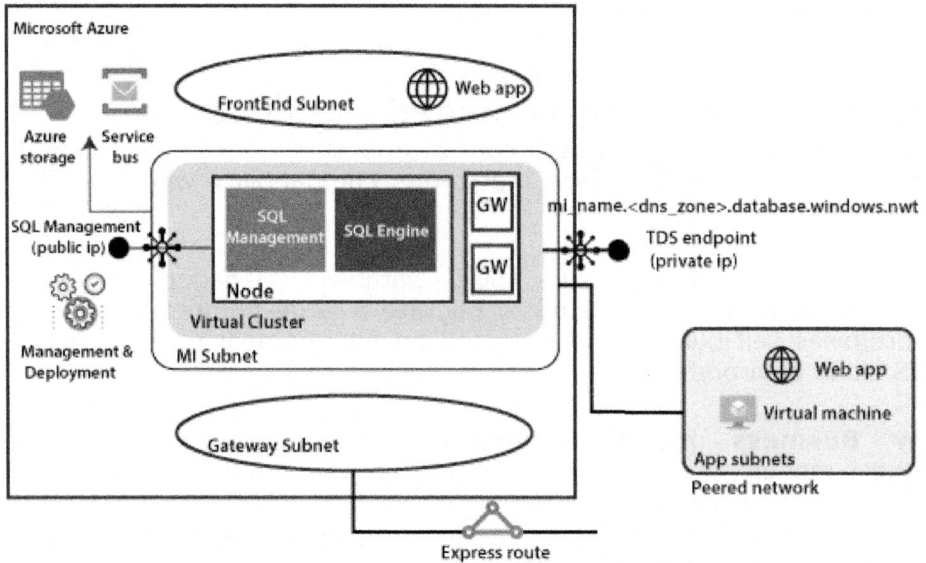

When we create a managed instance, a virtual network will get created. It will have front end subnet, Gateway subnet, a managed instance subnet, and the node that we can deploy as part of managed instance creation will get implemented into the MI subnet. Each node consists of the SQL engine and SQL management. Within the same network, we can deploy multiple nodes also, and these various nodes will form a virtual cluster with Gateway servers.

The entire virtual cluster will have two endpoints. The first endpoint will be for client connections, and the second endpoint is public but will be used by Microsoft to manage this environment. They need to connect to this environment using some automated script or something like that and maintain it for that purpose.

There is an endpoint also for this entire environment to work correctly. It needs to connect to Azure storage and service bus also. So, when we are trying to restrict the traffic from our MI subnet to the outside. Make sure we allow all the traffic related to Microsoft otherwise, our environment might not work correctly.

Finally, in terms of client connections and applications to connect to the database, they can reside in Frontend subnet and connect to the database, or they can live in a peered network when we peer the network which our MI subnet then all the web apps or virtual machines can be able to connect to the database because both networks have peered. We can also join our on-premises applications to the database by creating either a virtual network gateway or express gateway.

All the connections, whether it forms the web apps, or the virtual machines, or on-premises applications, all of them are communicating with the database over a private connection.

Azure COSMOS Database

Azure Cosmos DB is a NoSQL data store. It is different from the traditional relational database where we have a table, and the table will have a fixed number of columns, and each row in the table should adhere to the scheme of the table. In the NoSQL database, you don't define any schema at all for the table, and each item or row within the table can have different values, or different schema itself.

Advantages of Cosmos DB

❖ **No Schema & Index management:** The Azure database engine is fully schema-agnostic. Therefore no schema and index management are required. We also don't have to worry about application downtime while migrating schemas.

❖ **Industry-leading comprehensive SLAs:** Cosmos DB is the first and only service to offer industry-leading full 99.99% high availability, read and write latency at the 99th percentile, guaranteed throughput, and consistency.

❖ **The low total cost of Ownership:** Since Cosmos DB is a fully managed service, we no longer need to manage and operate complex multi-datacenter deployment, and upgrades of our database software pay for the support, licensing, or operations.

❖ **Developing application using NoSQL APIs:** Cosmos DB implements Cassandra, MongoDB, Gremlin, and Azure Table Storage wire protocol directly on the service.

❖ **Global distribution:** Cosmos DB allows us to add or remove any of the Azure regions to our Cosmos account at any time, with a click of a button.

Cosmos Database Structure

Database: We can create one or more Azure Cosmos database under our account. A database is analogous to a namespace, and it is the unit of management for a set of Azure Cosmos containers.

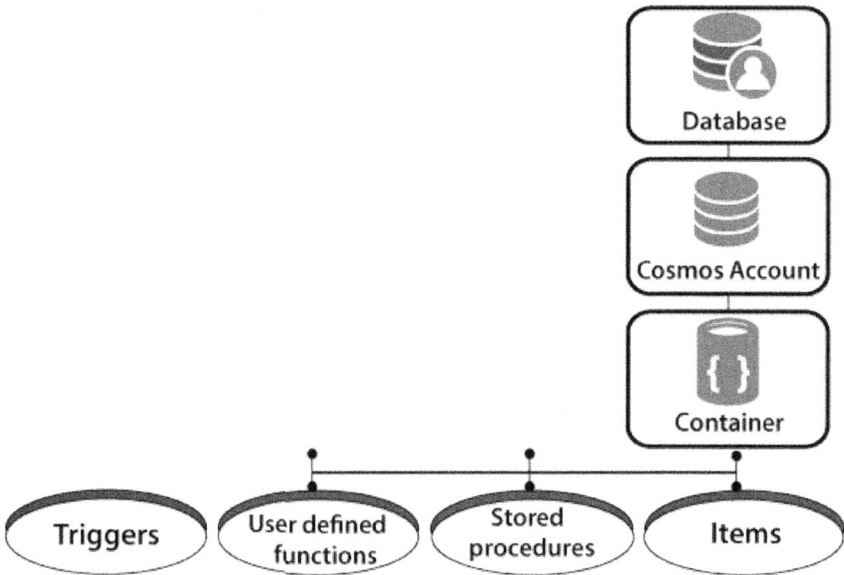

Cosmos Account: the Azure Cosmos account is the basic unit of global distribution and high availability. For globally distributing our data and throughput across multiple Azure regions, we can add or remove Azure regions from our Azure Cosmos at any time.

Container: An Azure Cosmos container is the unit of scalability for both provisioned throughput and storage of items. A container is horizontally partitioned and then replicated across multiple regions.

Global distribution and Partitioning

Cosmos DB works differently from the traditional relational database where we have a table, and all the rows in the table will sit in one physical place. When it comes to Cosmos DB, we will create logical partitions within a container so that we can have a certain amount of items with one partition key and a certain number of items with another partition key. They are called logical partitioning, and each logical partitioning can reside in a physical partition.

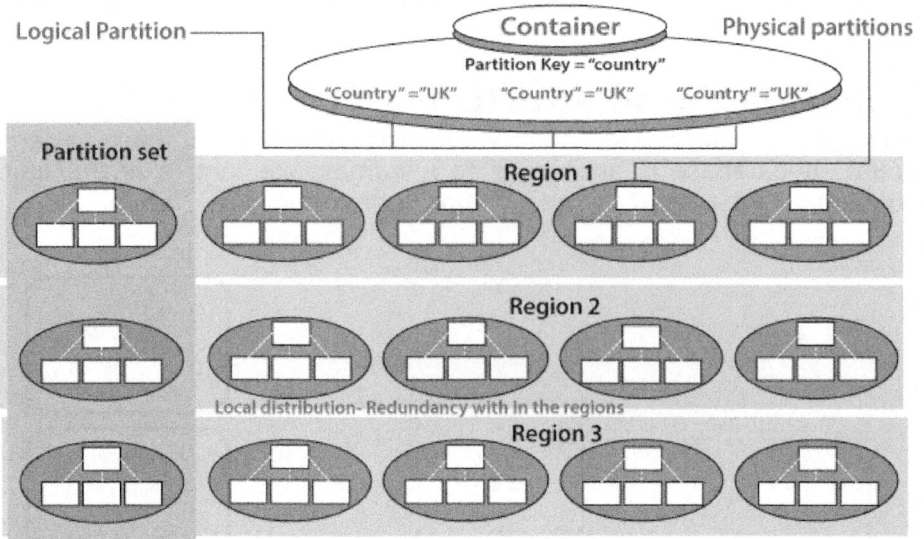

A container can contain millions of items, and we can divide these millions of items using partition key and make logical partitions, and each logical partition will reside in a physical partition. This is how the load of the container will be distributed across the board locally and also the data from here.

Types of Consistency

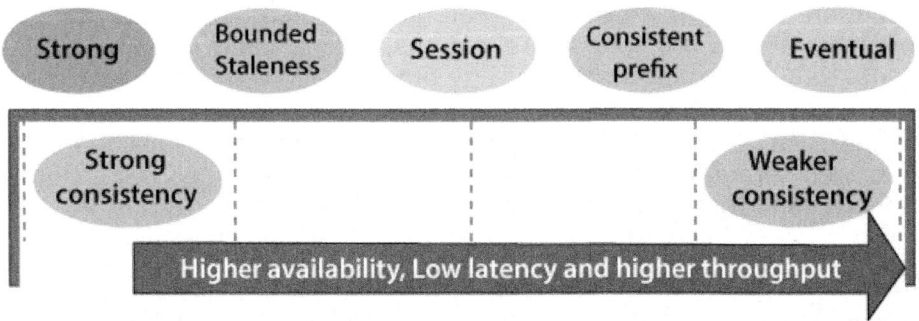

Azure Cosmos Database approaches the data consistency as a spectrum of choices instead of two extremes. Strong compatibility and eventual consistency are at the ends, but there are many consistency choices along the spectrum.

The consistency levels are region-agnostic. The consistency level of our Azure Cosmos account is guaranteed for all read operations regardless of the region from which the reads and writes are served, the number of areas associated with our Azure Cosmos account, or whether our account is configured with a single or multiple write regions.

Request Units

❖ We pay for the throughput we provision, and the storage we consume on an hourly basis with Azure Cosmos DB.
❖ The cost of all the database operations is normalized by Azure Cosmos DB and is expressed in terms of Request Units (RUs). The price to read a 1-KB item is 1 Request Unit (1 RU). All other database operations are similarly assigned with a cost in terms of RUs.
❖ The number of RU's consumed will depend on the type of operation, item size, data consistency, query patterns, etc.

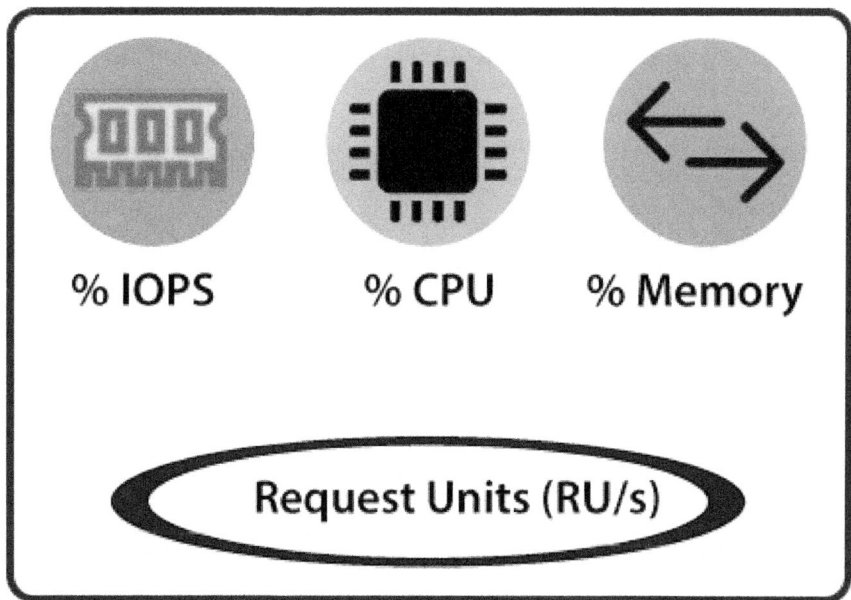

% IOPS   % CPU   % Memory

Request Units (RU/s)

For the management and planning of capacity, Azure Cosmos DB ensures that the number of RUs for a given database operation over a given dataset is deterministic.

Creating Azure Cosmos DB using Azure portal

Step 1: Click on create a resource and search for Azure Cosmos DB. After that, click on create.

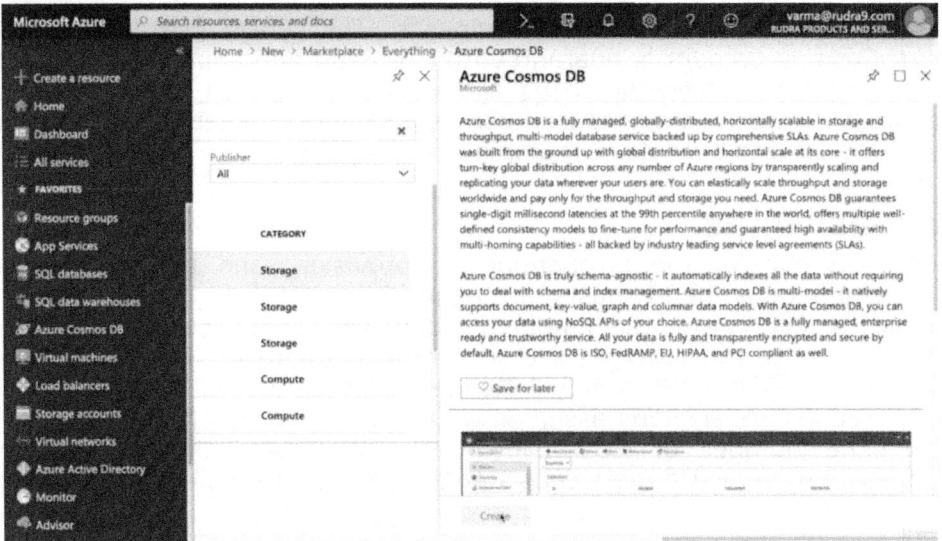

Step 2: Fill-in all the details and click on a review to see if any details are missing.

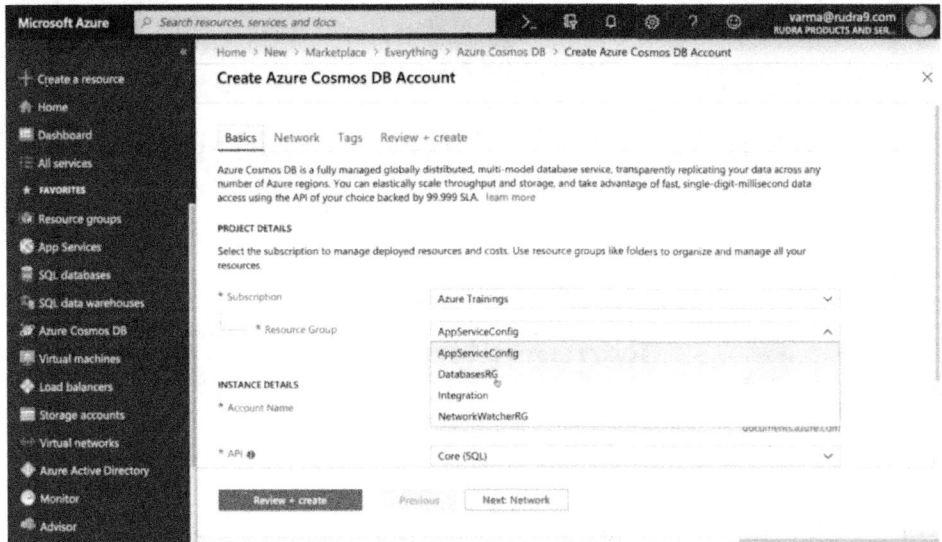

Step 3: Configure the network for the Azure Cosmos DB.

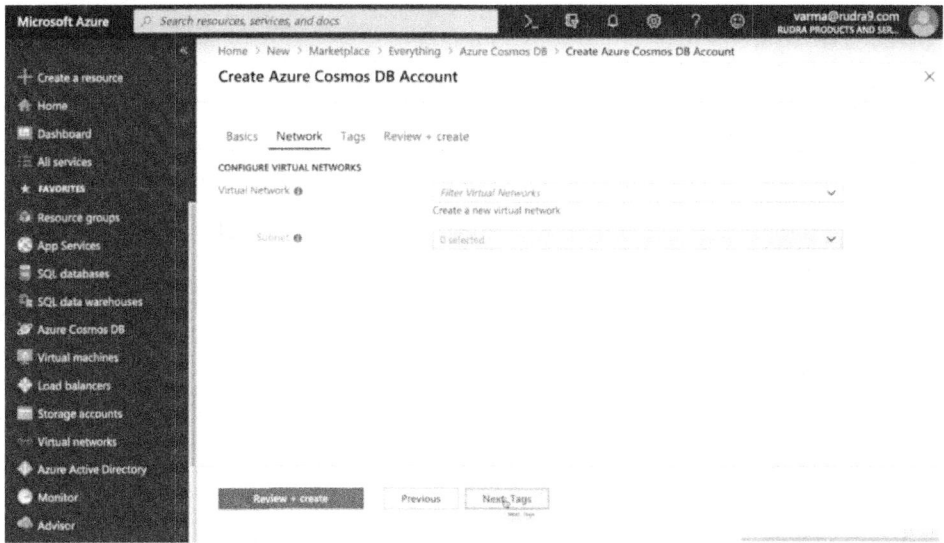

Step 4: Finally, click on the create button to create your COSMOS database.

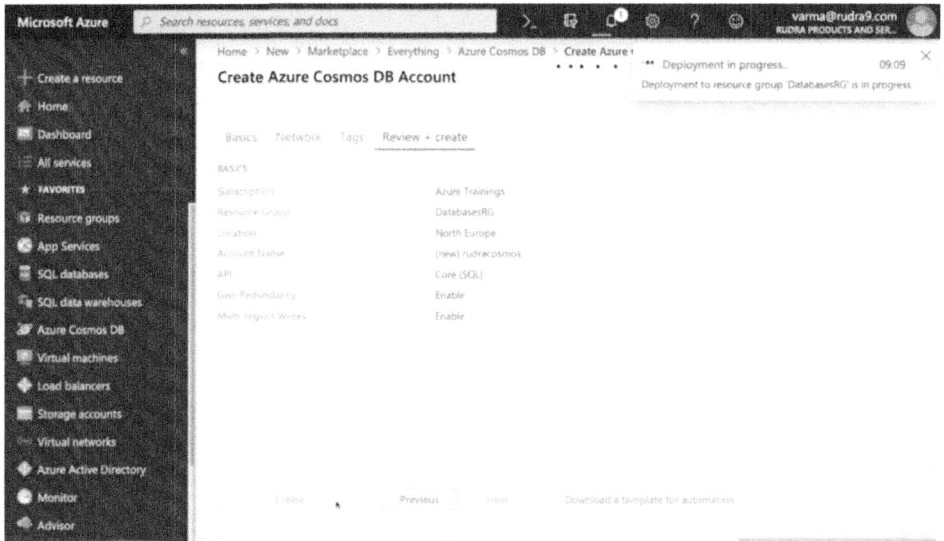

Step 6: You will see the following window after the successful deployment of Azure Cosmos DB.

Microsoft Azure

Search resources, services, and docs

varma@rudra9.com
RUDRA PRODUCTS AND SER...

- Create a resource
- Home
- Dashboard
- All services
- FAVORITES
- Resource groups
- App Services
- SQL databases
- SQL data warehouses
- Azure Cosmos DB
- Virtual machines
- Load balancers
- Storage accounts
- Virtual networks
- Azure Active Directory
- Monitor
- Advisor

Microsoft.Azure.CosmosDB-20190131090829 - Overview
Deployment

Search (Ctrl+/)

- Overview
- Inputs
- Outputs
- Template

🗑 Delete   ⊘ Cancel   ⬆ Redeploy   ↻ Refresh

⊘ Your deployment is complete

Go to resource

Deployment
name: Microsoft.Azure.CosmosDB-
20190131090829
Subscription: Azure Trainings
Resource group: DatabasesRG

DEPLOYMENT DETAILS (Download)
Start time: 31/01/2019, 09:09:32
Duration: 4 minutes 26 seconds
Correlation ID: cf714fd8-8014-4b97-9b8e-
9a0fb11b6efd

RESOURCE	TYPE	STATUS	OPERATI...
rudracosmo	Microsoft...	OK	Operation d

Additional
Resources

Windows
Server
2016 VM
Quickstart
tutorial

Cosmos
DB
Quickstart
tutorial

Web App
Quickstart
tutorial

SQL
Database
Quickstart
tutorial

Storage
Account
Quickstart
tutorial

[210]

Azure Data Factory

Azure Data Factory is a data-integration service based on the Cloud that allows us to create data-driven workflows in the cloud for orchestrating and automating data movement and data transformation. Data Factory is a perfect ETL tool on Cloud. Data Factory is designed to deliver extraction, transformation, and loading processes within the cloud. The ETL process generally involves four steps:

Connect & Collect 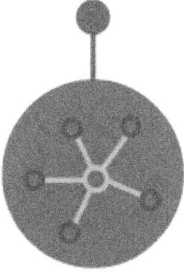 Publish

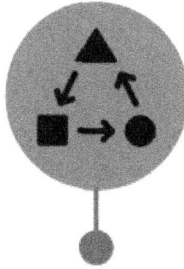 Transform  Monitor

- ❖ **Connect & Collect:** We can use the copy activity in a data pipeline to move data from both on-premises and cloud source data stores.
- ❖ **Transform:** Once the data is present in a centralized data store in the cloud, process or transform the collected data by using compute services such as HDInsight Hadoop, Spark, Data Lake Analytics, and Machine Learning.
- ❖ **Publish:** After the raw data is refined into a business-ready consumable form, it loads the data into Azure Data Warehouse, Azure SQL Database, and Azure Cosmos DB, etc.
- ❖ **Monitor:** Azure Data Factory has built-in support for pipeline monitoring via Azure Monitor, API, PowerShell, Log Analytics, and health panels on the Azure portal.

Components of Data Factory

Data Factory is composed of four key elements. All these components work together to provide the platform on which you can form a data-driven workflow with the structure to move and transform the data.

- ❖ **Pipeline:** A data factory can have one or more pipelines. It is a logical grouping of activities that perform a unit of work. The activities in a pipeline perform the task altogether. For example - a pipeline can contain a group of activities that ingests data from an Azure blob and then runs a Hive query on an HDInsight cluster to partition the data.

❖ **Activity:** It represents a processing step in a pipeline. For example - we might use a copy activity to copy data from one data store to another data store.

❖ **Datasets:** It represents data structures within the data stores, which point to or reference the data we want to use in our activities as I/O.

❖ **Linked Services:** It is like connection strings, which define the connection information needed for Data Factory to connect to external resources. A Linked service can be a data store and compute resource. Linked service can be a link to a data store, or a computer resource also.

❖ **Triggers:** It represents the unit of processing that determines when a pipeline execution needs to be disabled. We can also schedule these activities to be performed at some point in time, and we can use the trigger to disable an activity.

❖ **Control flow:** It is an orchestration of pipeline activities that include chaining activities in a sequence, branching, defining parameters at the pipeline level, and passing arguments while invoking the pipeline on-demand or from a trigger. We can use control flow to sequence certain activities and also define what parameters need to be passed for each of the activities.

Creating Azure Data-Factory using the Azure portal

❖ **Step 1:** Click on create a resource and search for Data Factory then click on create.

❖ **Step 2:** Provide a name for your data factory, select the resource group, and select the location where you want to deploy your data factory and the version.

❖ **Step 3:** After filling all the details, click on create.

❖ The Azure Data Factory completely had a different portal, as shown in the following figure.

Azure SQL Stretch Database & SQL Data Warehouse

SQL Stretch Database

It migrates our cold data transparently and securely to the Microsoft Azure Cloud. Stretch database divides the data into two types. One is the hot data, which is frequently accessed, and the second one is cold data, which is infrequently accessed. Also, we can define policies or criteria for hard data and cold data.

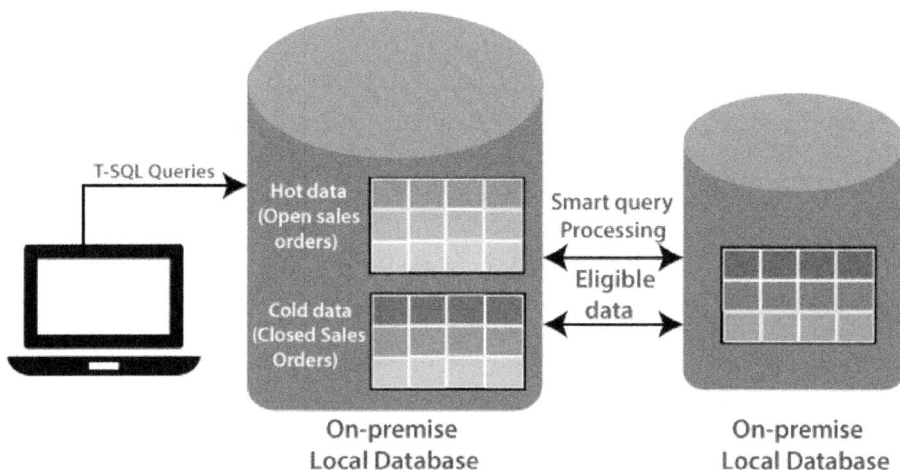

On-premise
Local Database

On-premise
Local Database

For example - if we have a sales order table, all those open and in Program Sales orders can be hot data, and all the closed sales orders can be cold data. The cold data will be transparently migrated to Azure SQL Stretch Database. However, it doesn't mean that we need to change our application in such a way that for open sales orders, we need to go to Azure SQL Stretch Database.

We can use the same queries in our application to fetch the data and based on the location of data, and the query will be automatically sent to Stretch Database.

Advantages of SQL Stretch Database

❖ It provides cost-effective availability for cold data that benefits from the low cost of Azure rather than scaling expensive on-premises storage.
❖ It doesn't require changes to the existing queries or applications. The position of the data is transparent to the application.

❖ It reduces the on-premises maintenance and storage for our data. Backups for our on-premises data run faster and finish within the maintenance window. Backups for the cloud portion of our data run automatically.

❖ It keeps our data secure even during migration. It provides encryption for our data in motion. Row-level security and other advanced SQL Server security feature also work with Stretch Database to protect our data.

SQL Data Warehouse

Microsoft SQL Data Warehouse within Azure is a cloud-based at scale-out database capable of processing massive volume of data, both relational and non-relational and SQL Data Warehouse is based on massively parallel processing architecture.

In this architecture, requests are received by the control node, optimized, and passed on to the compute nodes to do work in parallel. SQL data warehouse stores the data in Premium locally redundant storage, and linked to computing nodes for query extraction.

Components of SQL Data Warehouse

Data Warehouse units: Allocation of resources to our SQL Data Warehouse is measured in Data Warehouse Units (DWUs). DWUs is a measure of underlying resources like CPU, memory, IOPS, which are allocated to our SQL Data Warehouse.

Data Warehouse units provide a measure of three precise metrics that are highly correlated with data warehouse workload performance.

❖ **Scan/Aggregation:** Scan/Aggregation takes the standard data warehousing query. It scans a large number of rows and then performs a complex aggregation. It is an I/O and CPU intensive operation.

❖ **Load:** This metric measures the ability to ingest data into the service. This metric is designed to stress the network and CPU aspects of the service.

❖ **Create Table As Select (CTAS):** CTAS measures the ability to copy a table. It involves reading data from storage, distributing it across the nodes of the appliance, and writing it to storage again. It is a CPU, IO, and network-intensive operation.

Section VII: Azure DevOps

Azure DevOps provides developer service to support team to plan work, collaborate on code development, build and deploy the application.

For example - We have a very simple application, and the only developer can make changes to that application. Once the changes are completed, the application will be submitted to testing, and once the testing has been done successfully, it will be published into production. However, if our application is a very complex application with multiple modules, and we have different developers working on the enhancement of various modules within the application. Then it will become very complex to merge changes done by different developers and also take it through testing and finally building the application and deploying the application into production. The more developer we have, the more complicated the process is going to be; precisely that complexity can be addressed using Azure DevOps. We can use Azure DevOps to deploy both infrastructure and code into Azure.

Services of Azure DevOps

Azure DevOps has a number of services that we can take advantage of to manage our code development, building the application, deploying the applications, and also making our developers collaborate.

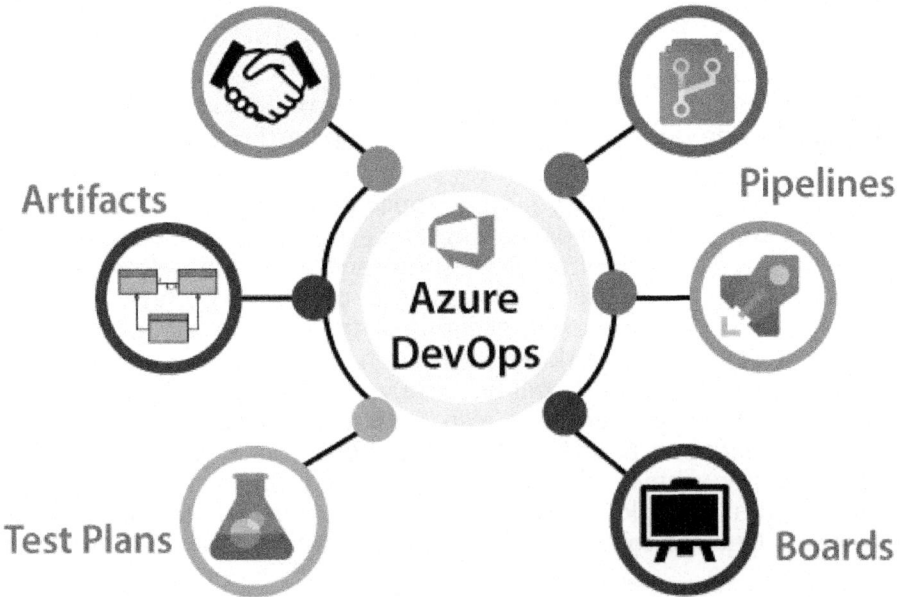

Collaboration tools   **Azure Repos**

Artifacts

Azure DevOps

Pipelines

Test Plans   **Boards**

❖ **Azure Repository:** It is a set of version control tools that we can use to manage our code. We can either use Git repositories or team foundation version control for source control of our code. In Azure repositories, we can create multiple branches, and each branch represents a version of code, and we can provide access to a particular branch to a specific developer.

❖ **Azure pipeline:** It is a fully-featured continuous integration and continuous delivery service. It works with our preferred Git provider and can deploy to most major cloud services, which includes Azure services also. Using Azure pipelines, we can able to define a build pipeline to build our code and also a release pipeline to carry out release into a specific destination.

❖ **Boards:** It provides a rich set of capabilities, including native support for Scrum and Kanban, customizable dashboards, and integrated reporting. We can create different activities, track activities, and we can move activities between different buckets like Dun bucket, backlog bucket, in progress bucket, etc.

❖ **Test Plan:** It provides a browser-based test management solution with all capabilities required for planned manual testing, exploratory testing, etc.

❖ **Artifacts:** It is very important because most of our application will have some dependency on different packages, for example, NuGet package, npm, Maven package, etc. It also supports universal Packages, which can store any file or set of files.

❖ **Collaboration tools:** It includes a customizable team dashboards with configurable widgets to share information progress and trends. We can create Wiki packages for sharing information, and also we can configure some notifications.

Structure of DevOps

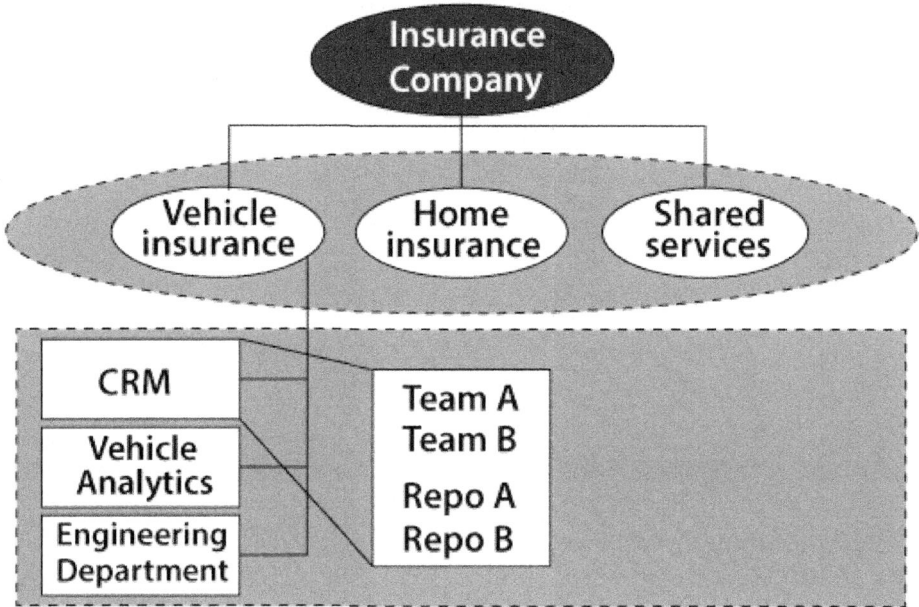

Organization: An organization in Azure DevOps is a mechanism for organizing and connecting groups of related projects. For example - business divisions, regional divisions, or other enterprise structures.

Projects: A project contains a following set of features in Azure DevOps:

❖ Boards and backlogs for agile planning
❖ Pipelines for continuous integration and deployment.
❖ It contains repositories for version control, management of source code, and artifacts.
❖ It keeps continuous test integration throughout the project life cycle.

Azure DevOps Portal

Azure DevOps portal is a centralized portal where we can manage all the Azure DevOps services. We need to create an account on the Azure portal to avail of all of the facilities. For the training purpose, we will take the free services. To create an Azure portal, organization, and project follow these steps carefully.

Step   1: Go   to https://Azure.microsoft.com/en-in/services/devops/ and click on **Start Free**.

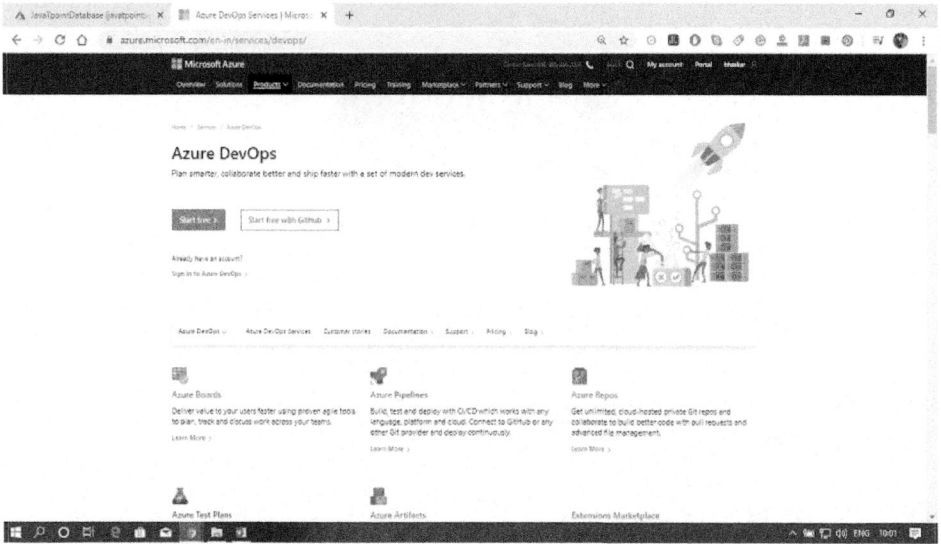

Step 2: After that, it will ask you to fill the details and region. Fill the details and click next.

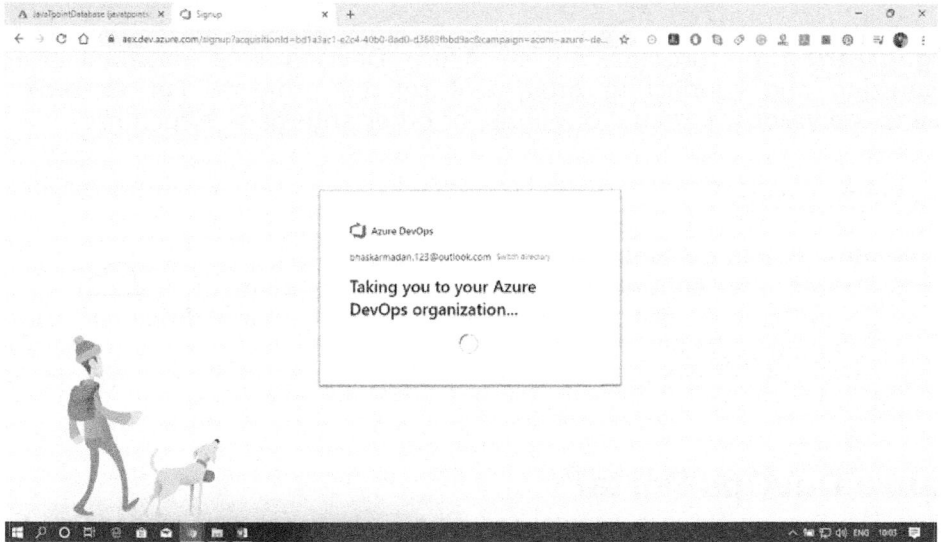

Step 3: Your Azure DevOps account has been created. And now, we will create an organization and create a project inside the organization.

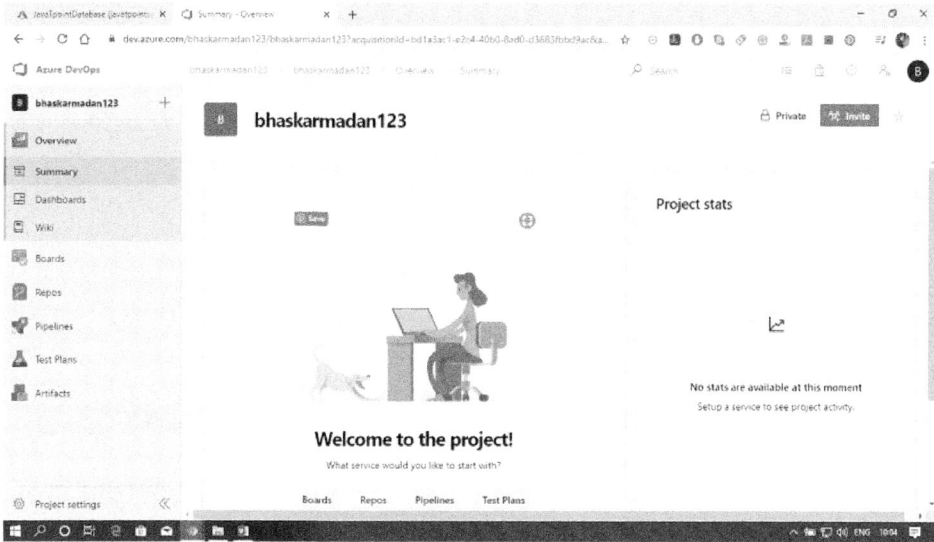

Step 4: Click on *New Organization*.

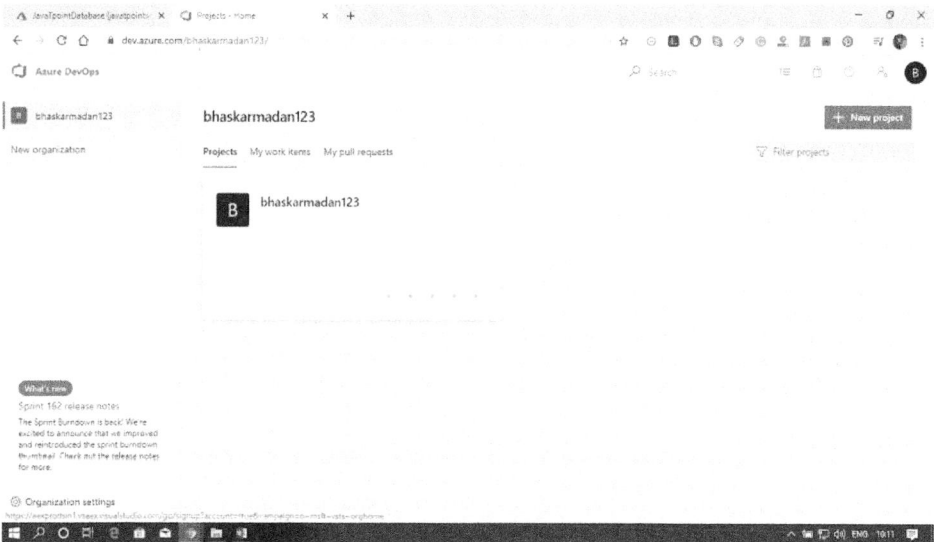

Step 5: Provide a name to your organization and select the location from where you want to get hosted your organization.

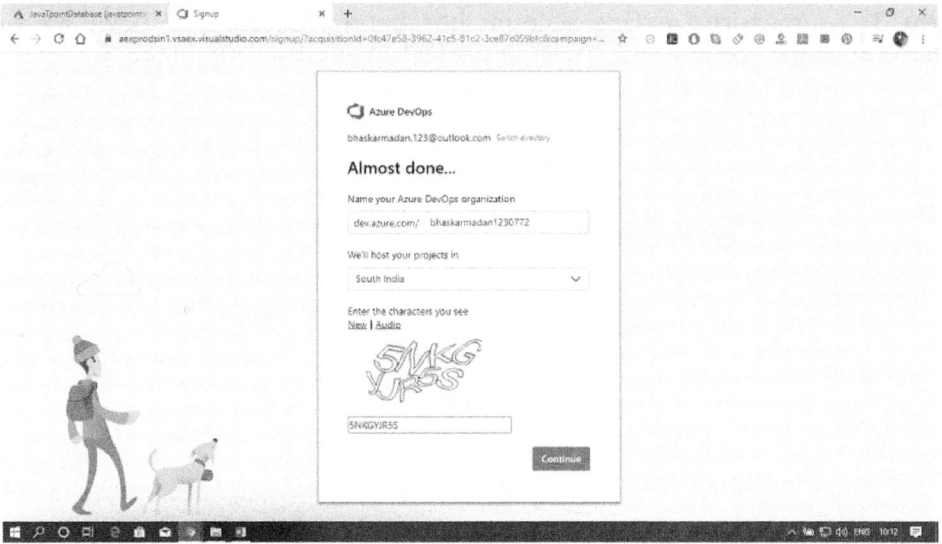

Step 6: Your organization has been created. Here you will see the Create a project page.

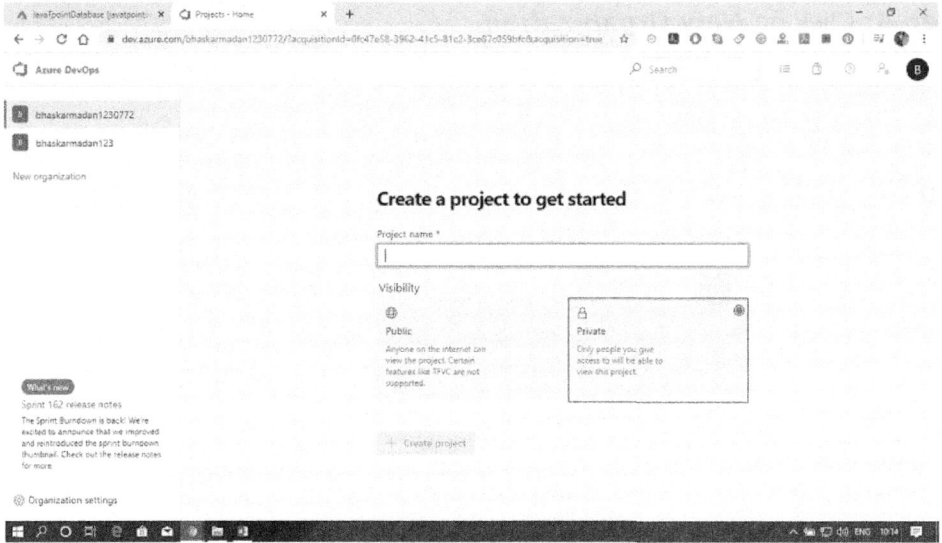

Step 7: Provide a name to your project. After that, click on *advanced,* then select the version control and work item process.

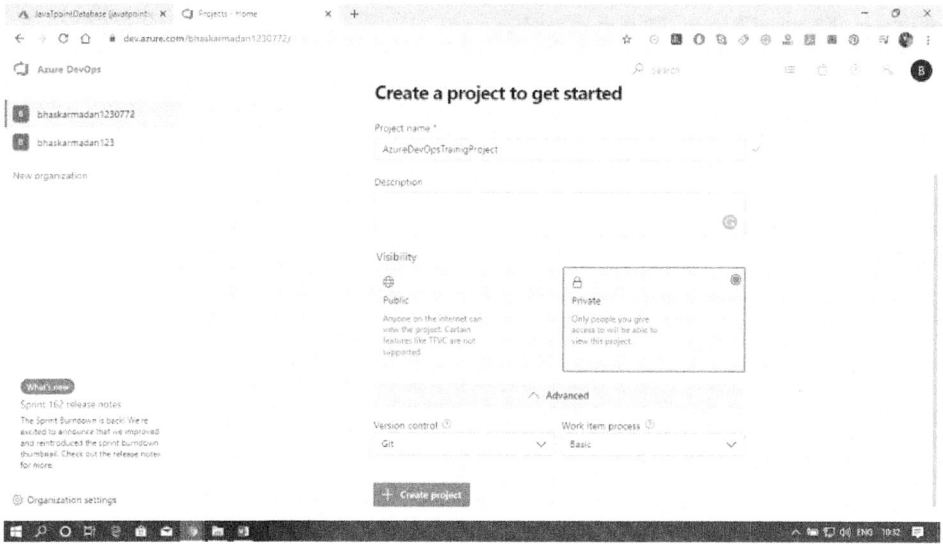

Step 8: Finally, click on the *create project* button. Your project has been created. Now, you can invite members to your project.

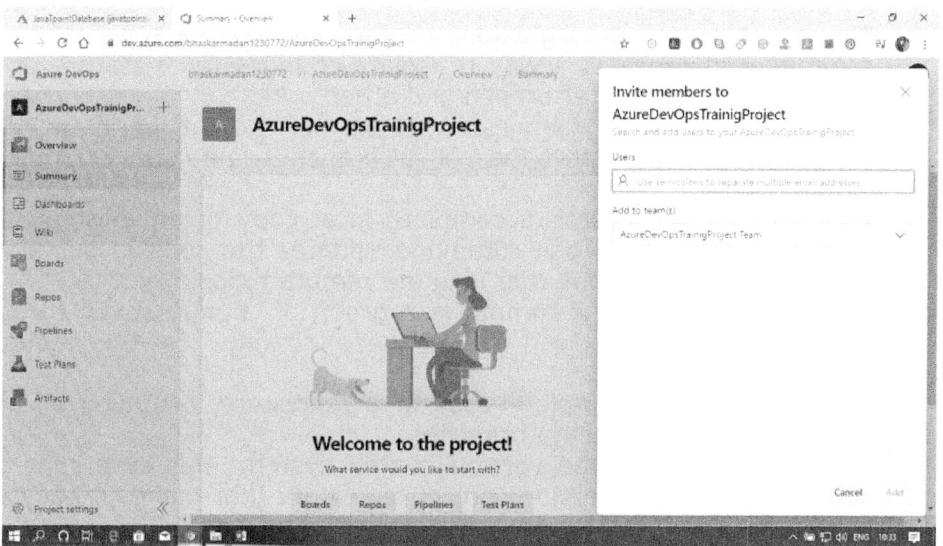

Azure DevOps Repository

Azure Repository is a set of version control tools that we can use to manage our code. In case if we are entirely new to version control, then version control enables us to track changes we make in our code over time. There are so many software that is available in the market to enable version control on our code. We can use the version control system to keep track of each change done by each developer, safely merge them, test the changes, and publish the change into production.

There are two types of version control in Azure Repos.

- ❖ **Git:** It is a distributed version control.
- ❖ **Team Foundation Version Control:** It is a centralized version control.

Azure Repos Concepts

1. **Repository:** A repository is a location for our code, which is managed by version control. It supports Git and TFVC so we can create multiple repositories in a single project and various branches for each repository.
2. **Branch:** A branch is a lightweight reference that keeps a history of commits and provides a way to isolate changes for a feature or a bug fix from our master branch and other work.
3. **Branch policies:** It is an essential part of the Git workflow. We use them to help protect the critical branches is our development, as the master.
4. **Pull and Clone:** Create a complete local copy of an existing Git repo by cloning it. A pull command updates the code in our local repository with the code that is in the remote repository.
5. **Push and Commit:** A commit is a group of change saved to our local repository. We can share these changes to the remote repository by pushing.
6. **Fork:** A fork is a complete copy of a repository, including all file commits, and (optionally) branches.
7. **Git:** Git is a distributed version control system. Our local copy of code is a complete version control repository that makes it easy to work offline or remotely.
8. **Notification:** Using notification, we will receive an email whenever any changes occur to work items, code reviews, pull requests, source control files and builds.
9. **Projects:** A project provides a place where a group of people can plan, track progress, and collaborate on building software solutions.
10. **Teams:** A team corresponds to a selected set of project members. With teams, organizations can subcategorize work to better focus on all of the work they track within a project.

Publish ARM Deployment project into DevOps

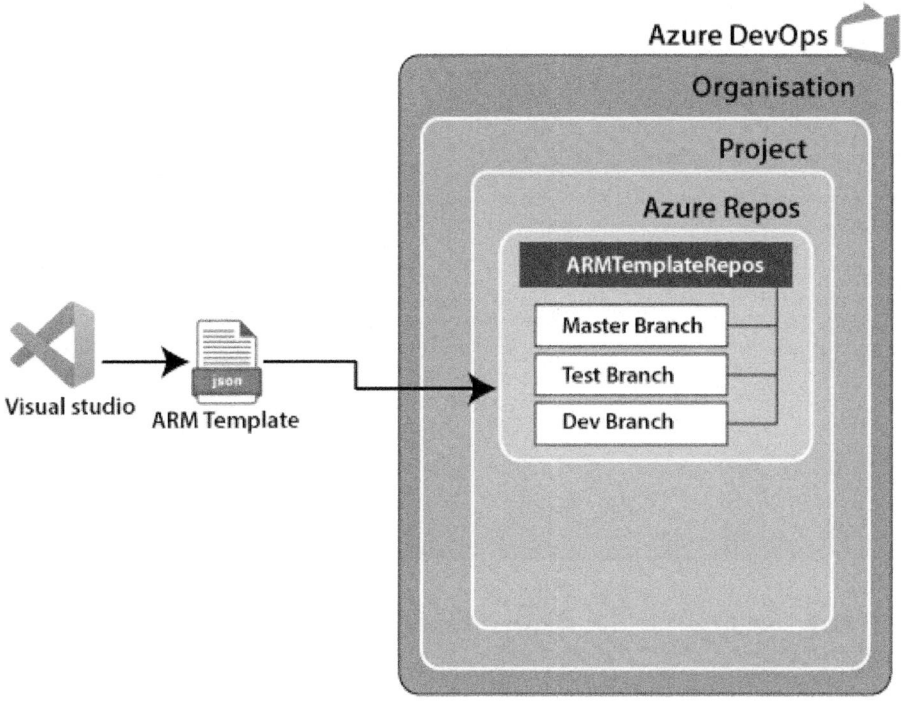

Step 1: Open Visual Studio and click on **create a new project**.

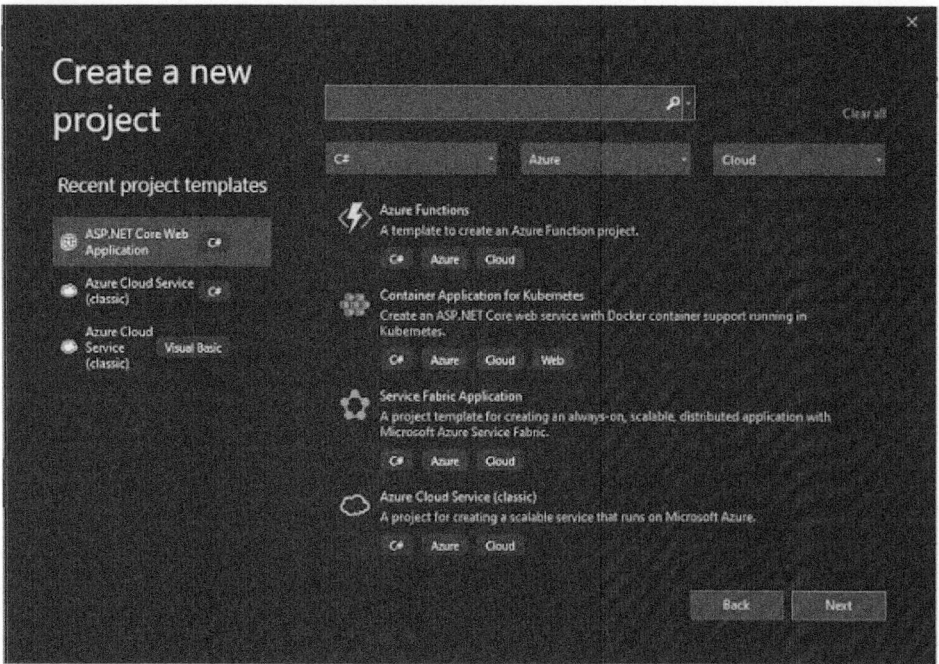

Step 2: Now, search for the *Azure Resource Group*. Then select it and click on Next.

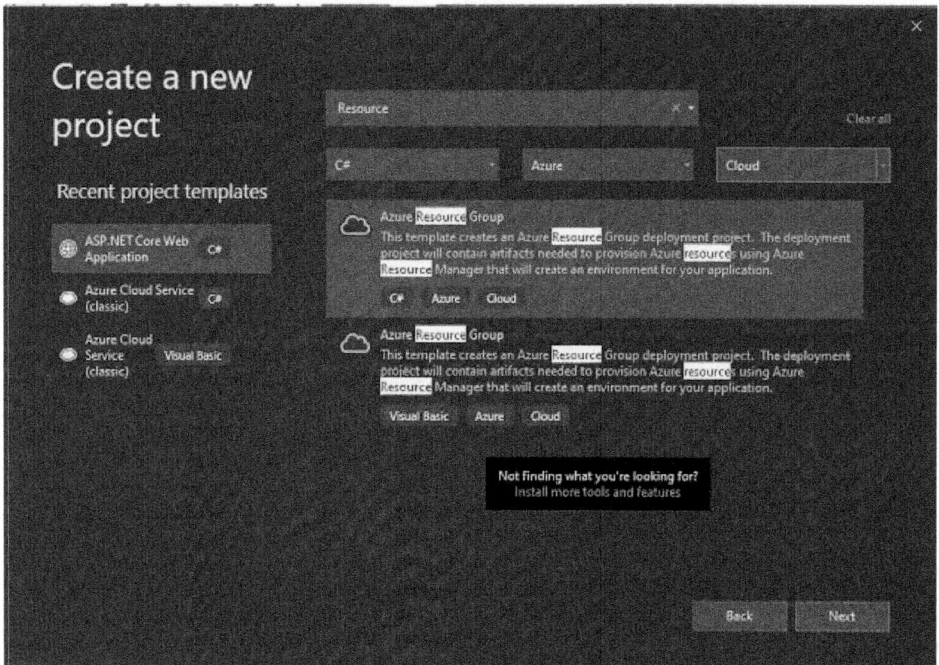

Step 3: Configure your new project and click on *create*.

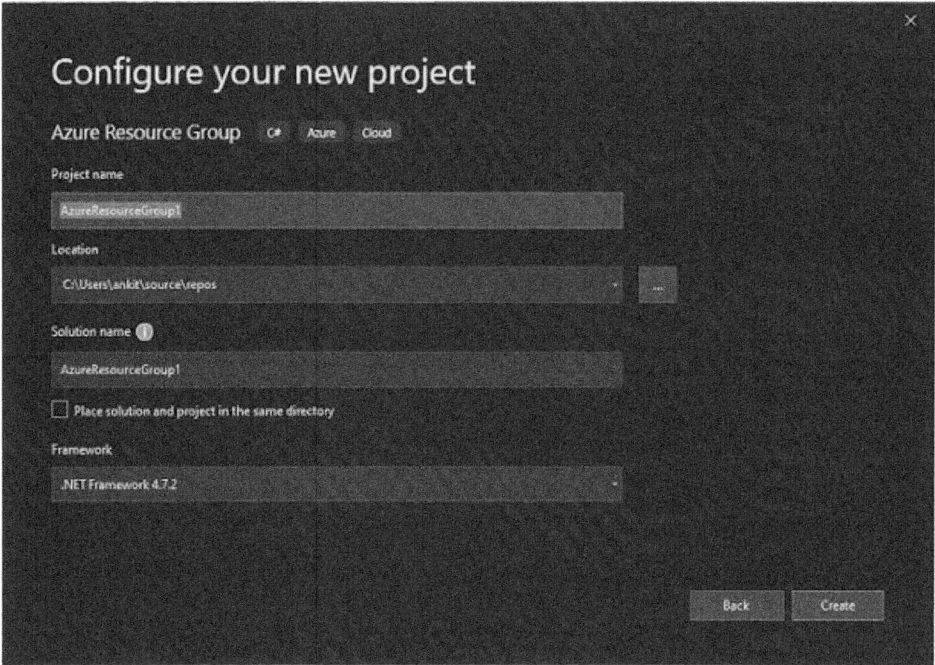

Step 4: Select the **Web app** from the available Azure Template in Visual Studio.

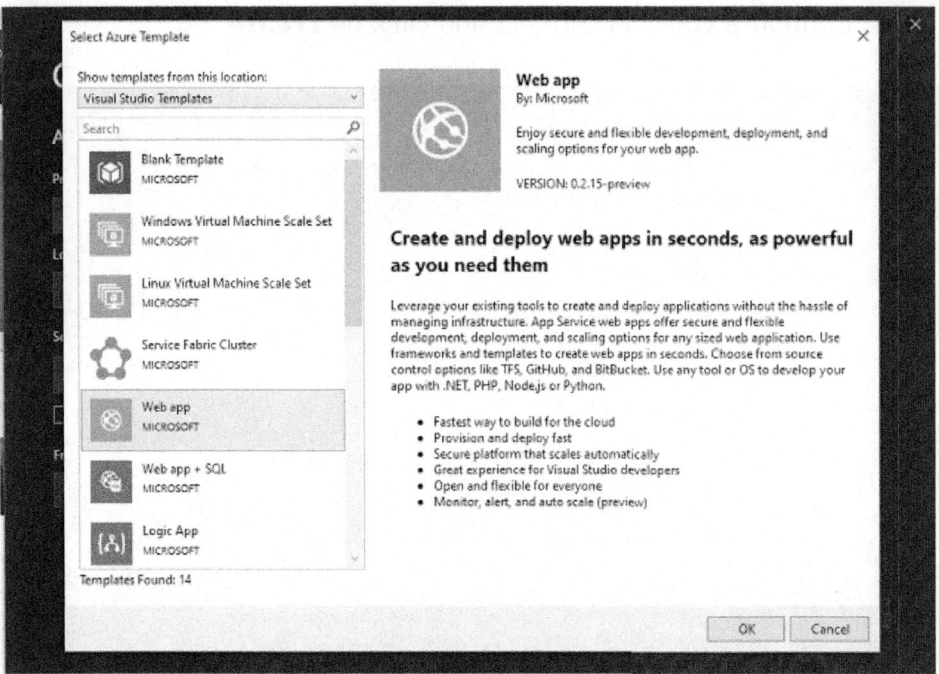

Step 5: Click on the *website.json file,* then you will find the available resources on the left-hand side file explorer.

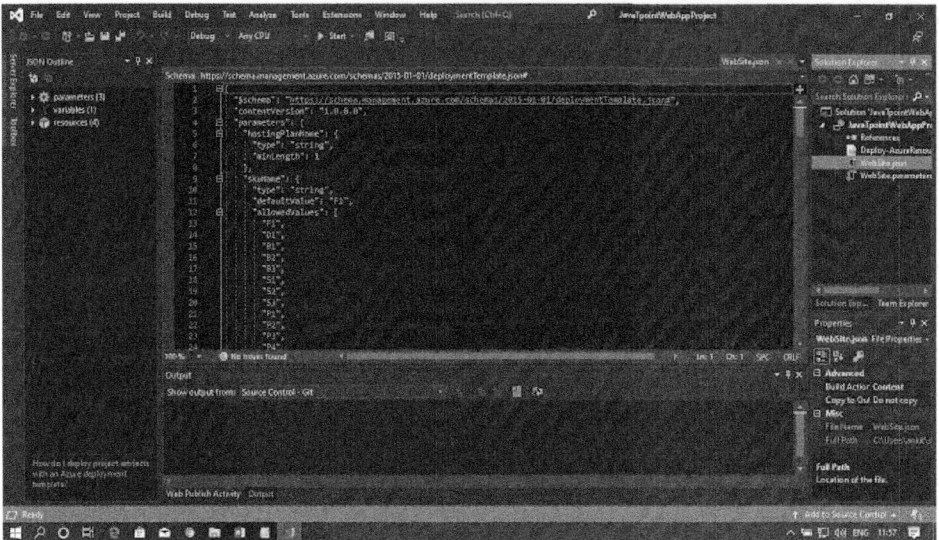

Step 6: Now, delete the *'appinsightcomponent'* resource from the file. Because we don't need this service right now.

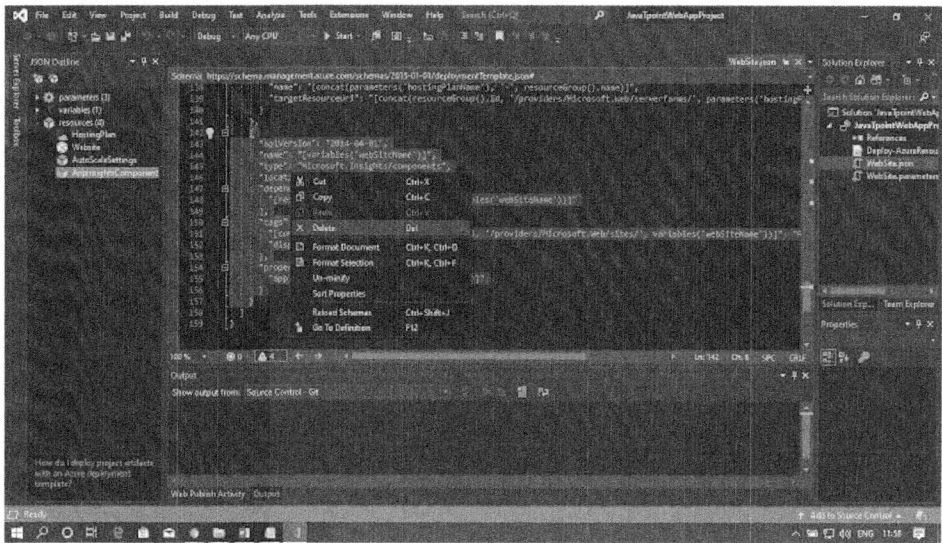

Step 7: To publish this code to the Azure DevOps portal, add this solution to source control. Then Right-click on the solution and then click on **add solution to source control**.

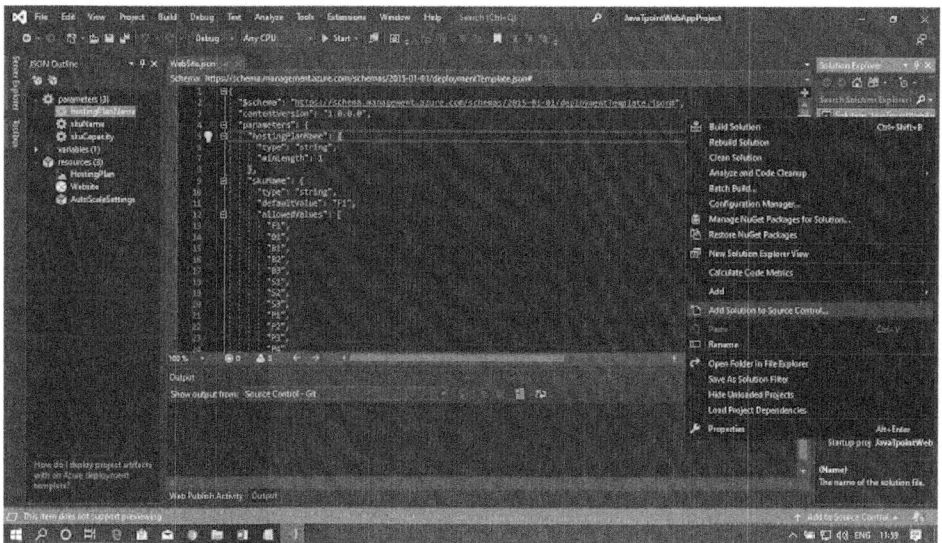

Step 8: A new git repository has been created. You can see the message in the output window.

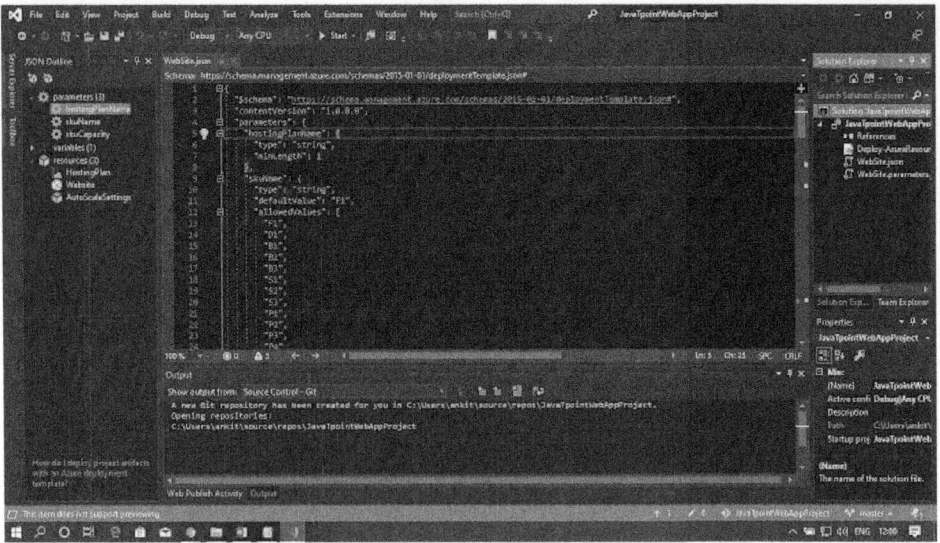

Step 9: Click on **Team Explorer** as shown in the figure below, then right-click on the dropdown menu and select sync.

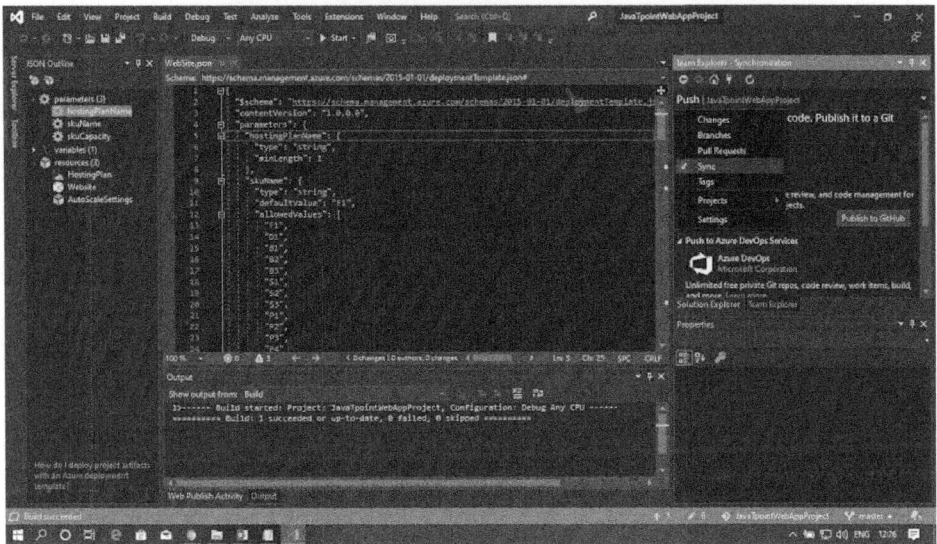

Step 10: Now, Click on the **Publish Git Repo** button to publish this project in the Azure DevOps organization.

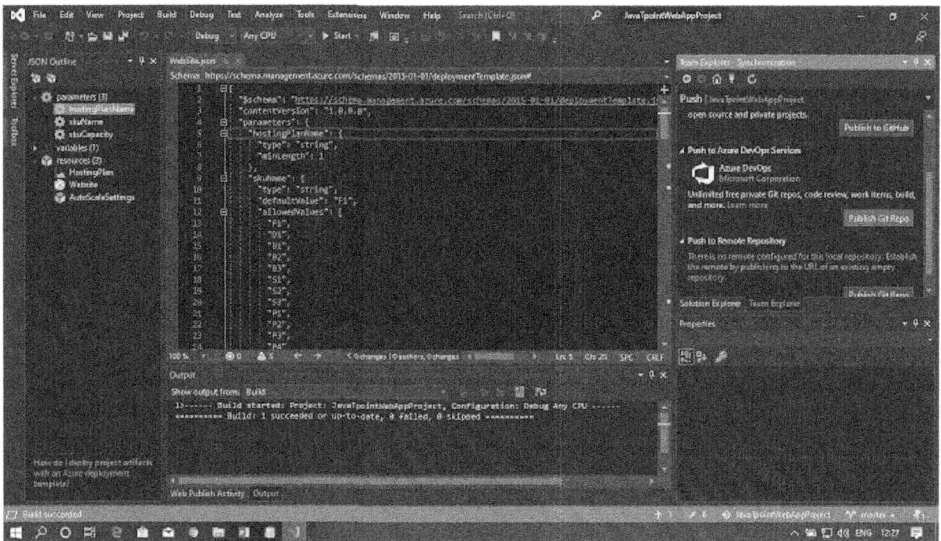

Step 11: Select the project and repository where you want to push this git repository in the Azure DevOps portal. Finally, Click on **publish repository**.

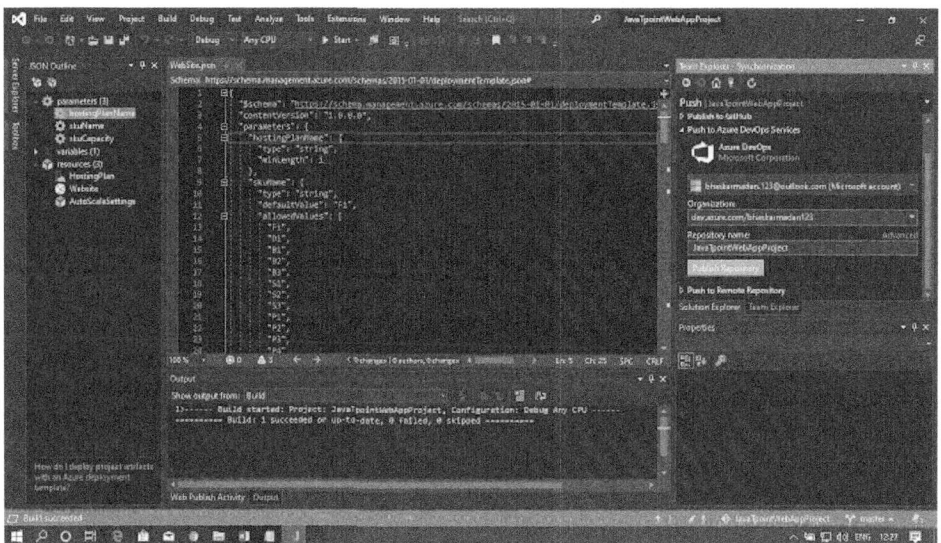

Step 12: To see your repository, open the Azure DevOps portal. And go through the organization that you have selected during the publishing. Click on the Repos to view the files.

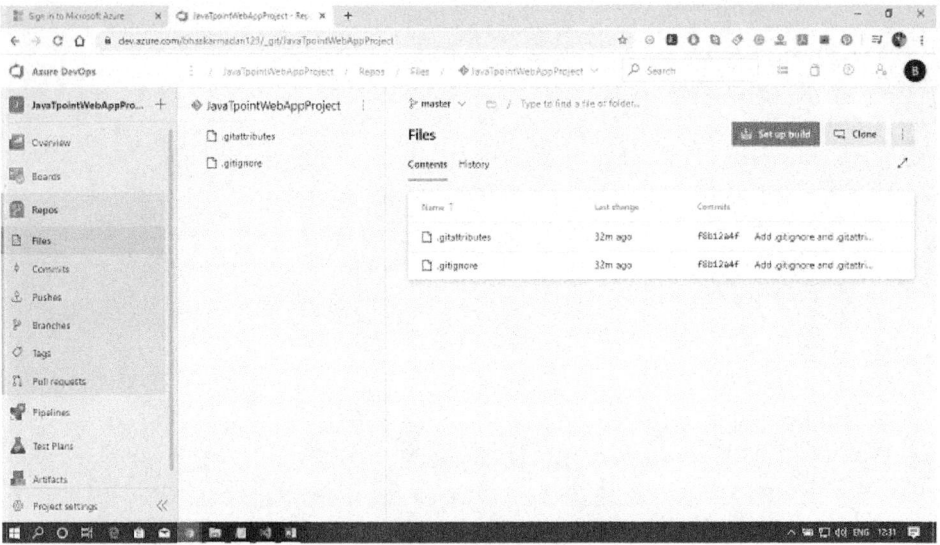

Step 13: To see the branches associated with your repository, click on Branches. Here we have only one branch right now, which is the default master branch.

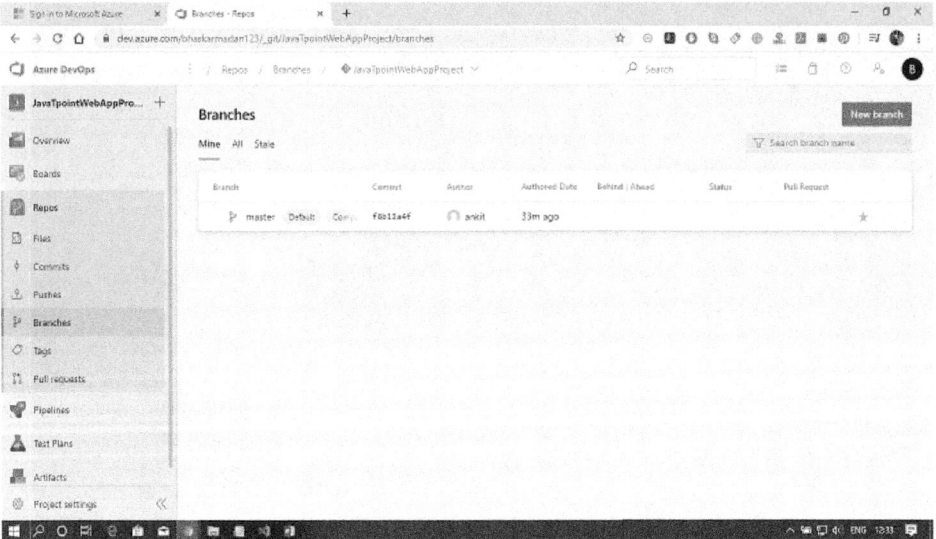

Azure DevOps Pipeline

Azure Pipeline is a cloud service that we can use to build and test our code project automatically. The Azure pipeline has a lot of capabilities such as continuous integration and continuous delivery to regularly and consistently test and builds our code and ship to any target.

There are three key distinct advantages of using Azure DevOps pipelines.

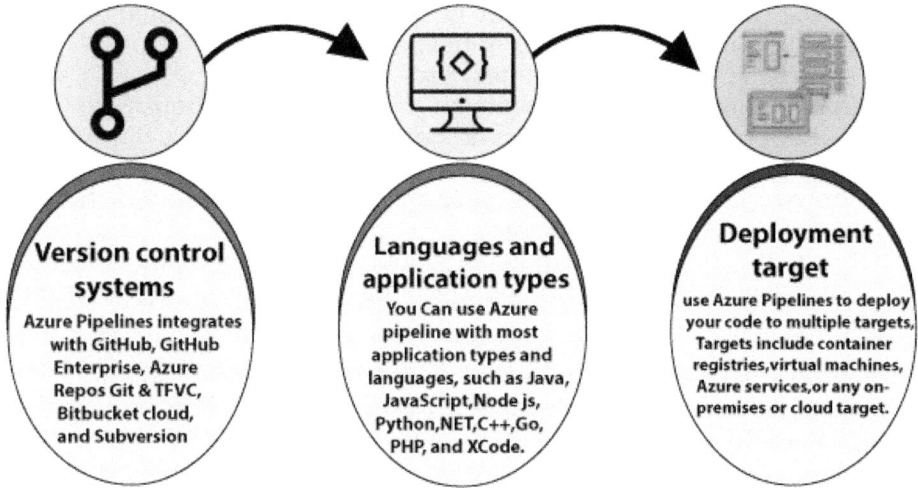

Version control systems

Azure Pipelines integrates with GitHub, GitHub Enterprise, Azure Repos Git & TFVC, Bitbucket cloud, and Subversion

Languages and application types

You Can use Azure pipeline with most application types and languages, such as Java, JavaScript,Node js, Python,NET,C++,Go, PHP, and XCode.

Deployment target

use Azure Pipelines to deploy your code to multiple targets, Targets include container registries,virtual machines, Azure services,or any on-premises or cloud target.

Version control system: Azure Pipelines integrates with GitHub, GitHub Enterprise, Azure Repos Git & TFVC, Bitbucket Cloud, and Subversion.

Language and application types: We can use Azure Pipeline with most application types and languages, such as Java, JavaScript, Node.js, Python, .Net, C++, Go, PHP, and Xcode.

Deployment target: We can use Azure Pipelines to deploy our code to multiple targets. Targets include - container registries, virtual machines, Azure services, or any on-premises or cloud target.

Azure DevOps Pipeline concepts
1. **Pipeline:** It is a workflow that defines how our test, build, and deployment steps are run.
2. **Stage:** It is a logical boundary in the pipeline. It can be used to mark the separation of concerns. Each stage contains one or more jobs.

3. **Job:** A stage can contain one or more jobs. Each job runs on an agent. It represents an execution boundary of a set of steps.
4. **Step:** It is the smallest building block of a pipeline. It can either be a script or a task. A task is simply an already created script offered as a convenience to you.
5. **Agent and Agent pools:** An agent is an installable software that runs one job at a time. Instead of managing each agent individually, you organize agents into agent pools.
6. **Artifact:** It is a collection of files or packages published by a run. The Artifact is made available to subsequent tasks, such as distribution or deployment.
7. **Trigger:** It is something that is set up to tell the pipeline when to run. We can configure a pipeline to run upon a push to the repository, at scheduled times, etc.
8. **Environment:** It is a collection of resources, where you deploy your application. It contains one or more virtual machines, containers, web apps, etc.
9. **Checks:** Checks define a set of validations required before a deployment can be performed.
10. **Runs:** It represents a single execution of a pipeline and collects the logs associated with running the steps and the results of running tests.

Publish ARM deployment project into DevOps Repos and deploy using pipeline

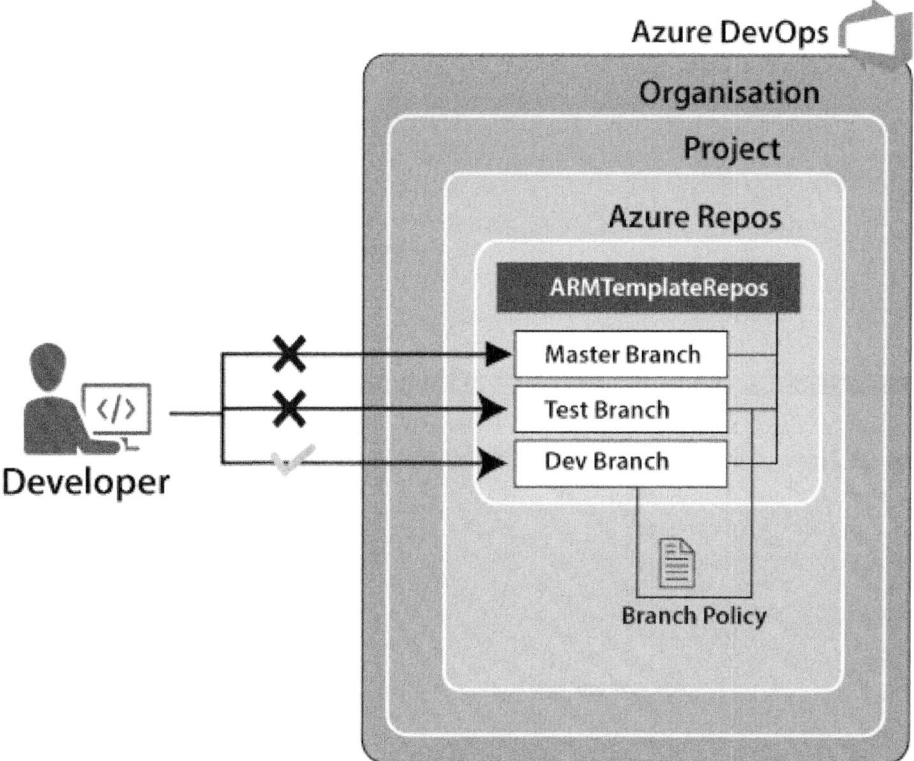

Step 1: Go into the Azure DevOps project and click on pipelines. After that, click on the New pipeline button.

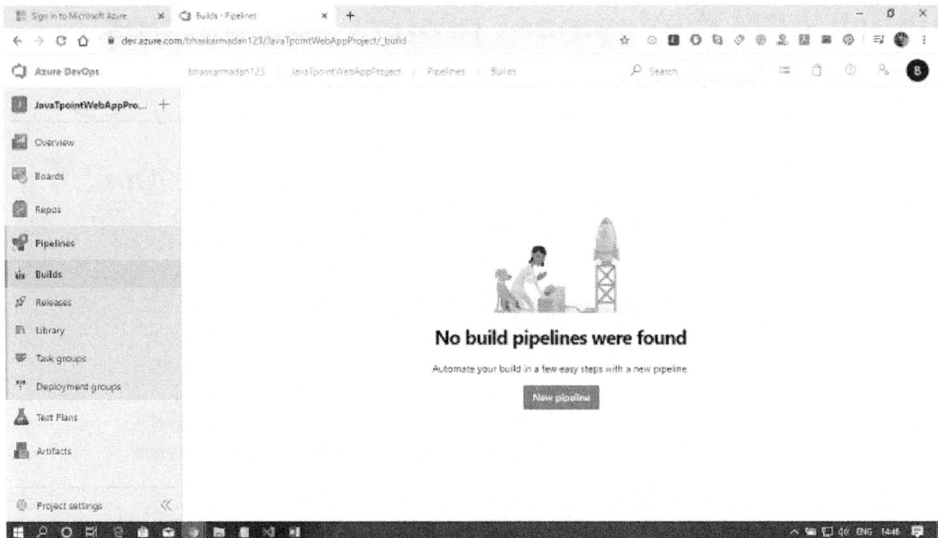

Step 2: Now, Click on the *"use the classic editor"* link down below.

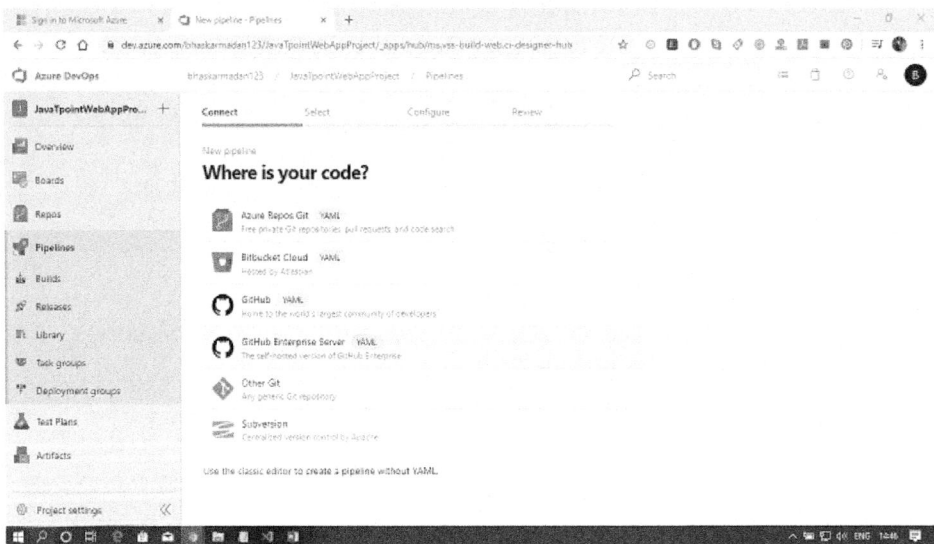

Step 3: Select the **project** and **repository** where you want to create the pipeline then click on **Continue**.

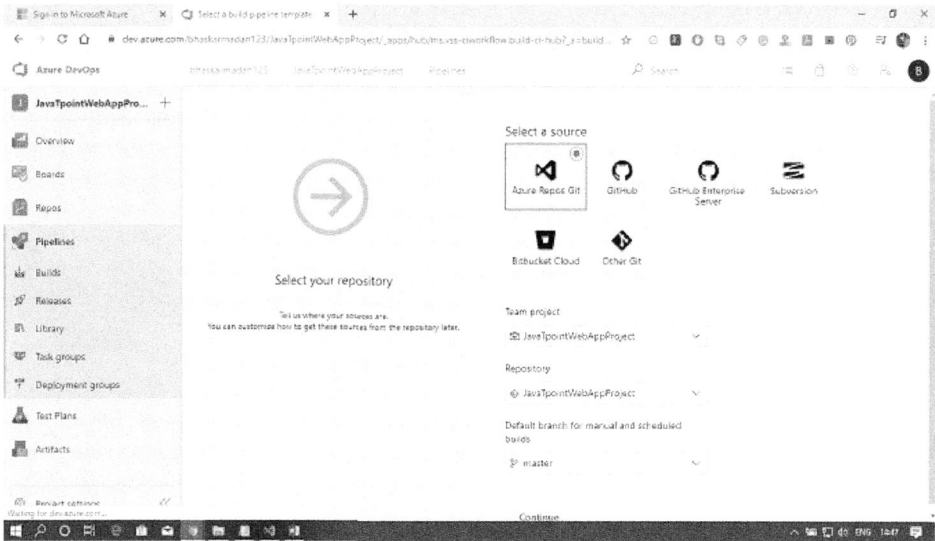

Step 4: Click on the **Empty job** link to create a job.

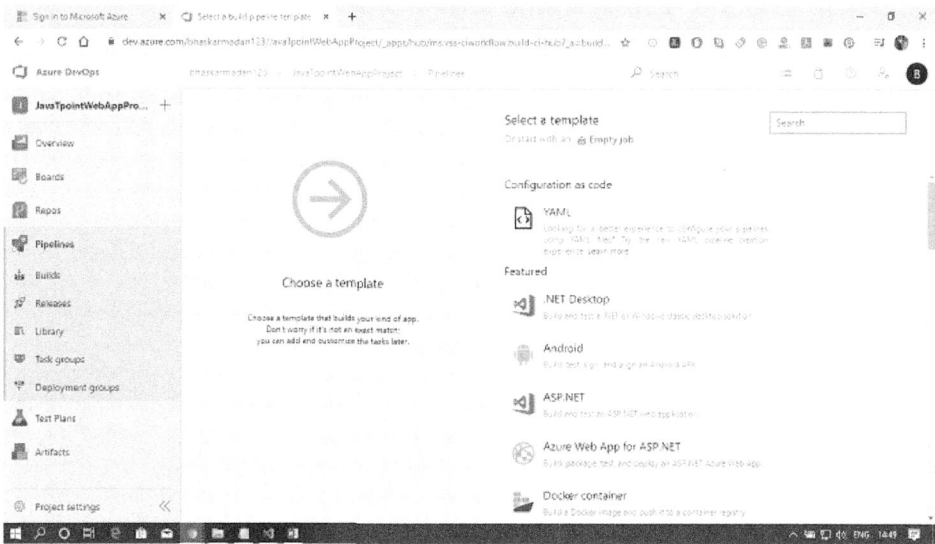

Step 5: Now, you need to add a task for building the activity. Click on the add button on the Agent job 1, then type-in resource group. Finally, click on the Azure Resource group deployment **add** button.

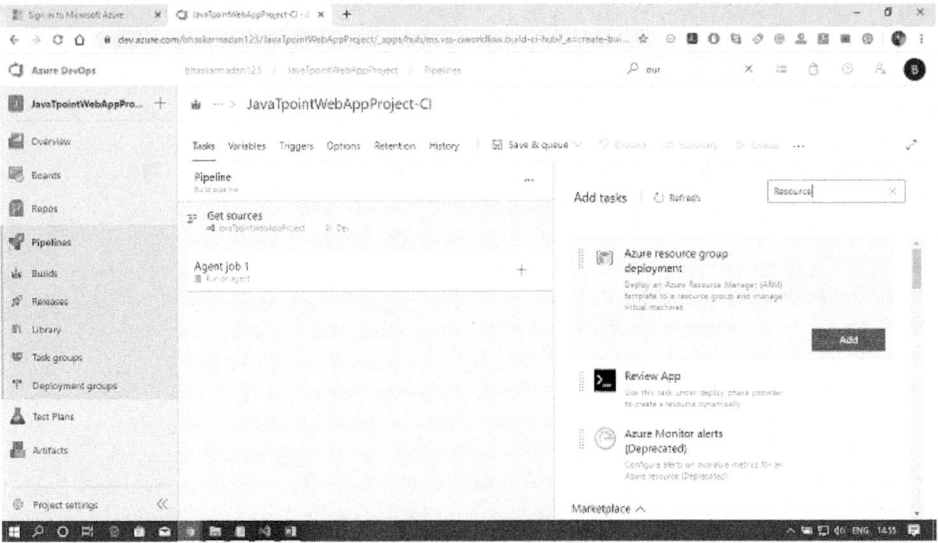

Step 6: Now, you need to select in which Azure subscription you want to deploy the infrastructure, into which resource group you want to deploy, and what you want to deploy in the form of JSON.

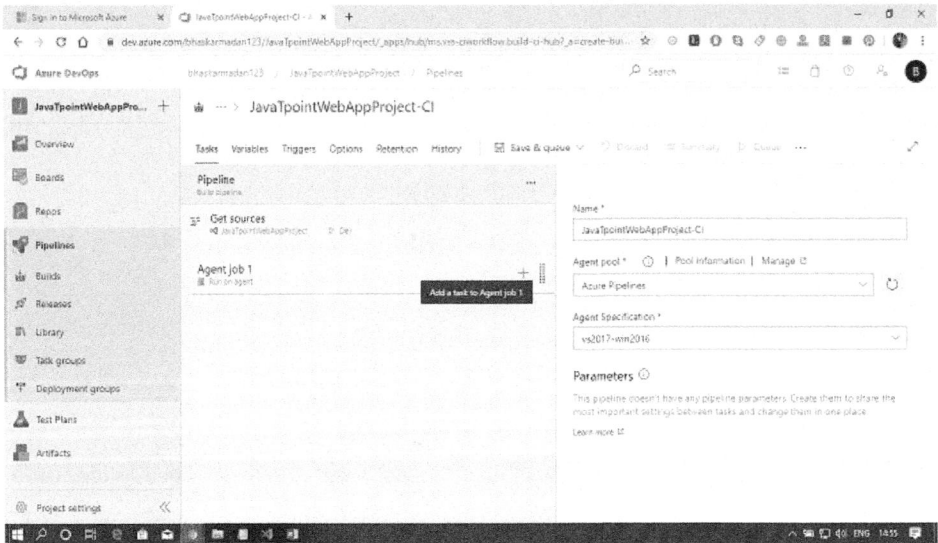

Step 7: Select the template from the repository.

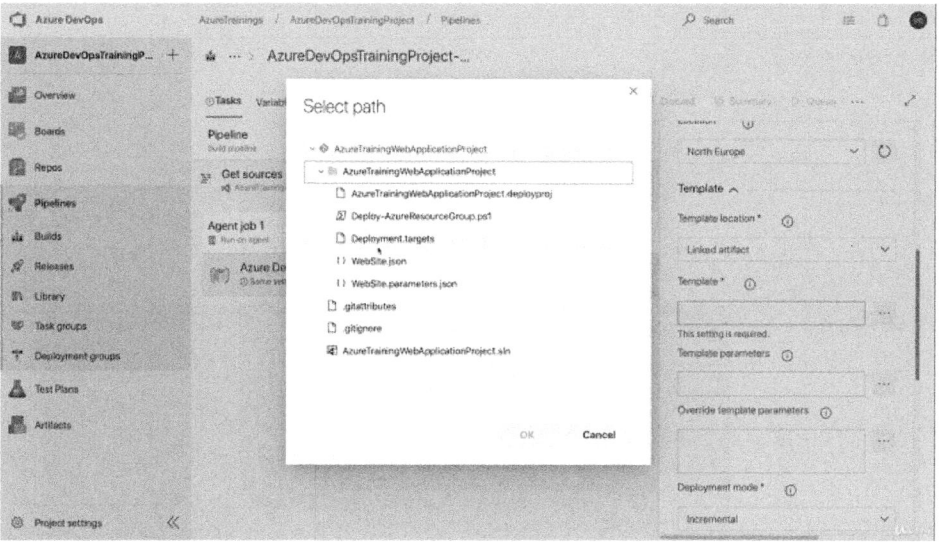

Step 8: After that, select the parameters file.

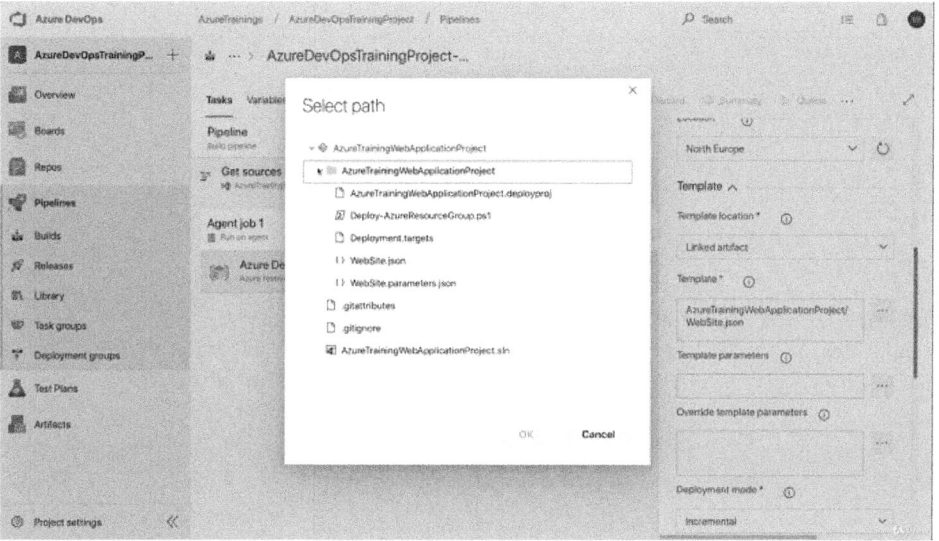

Step 9: Finally, click on Save & Queue. The deployment will take some time.

Azure DevOps

: / AzureTrainings / Pipelines / Builds / AzureDevOpsTrainingProject... / #10

Search

AzureDevOpsTrainingP... +

Overview

Boards

Repos

Pipelines

Builds

Releases

Library

Task groups

Deployment groups

Test Plans

Artifacts

Project settings

#10: PLan name has been given

Cancel build

Manually run just now by V S Varma Rudra raju @ AzureTrainingWebApplicationProject |> Dev ◇ ddb959b

Logs   Summary   Tests

Agent job 1
Pool: Hosted Windows 2019 with V... · Agent: Hosted Agent

Started: 08/08/2019, 09:58:11
5s

Initialize job · succeeded   1s

Checkout · succeeded   3s

Azure Deployment:Create Or Update Resource Group action on Training-Dev-RG   <1s

```
*********************************************************************************
Starting: Azure Deployment:Create Or Update Resource Group action on Training-Dev-RG
*********************************************************************************
==============================================================================
Task        : Azure resource group deployment
Description : Deploy an Azure Resource Manager (ARM) template to a resource group and manage virtual machine
s
Version     : 2.156.0
Author      : Microsoft Corporation
Help        : https://docs.microsoft.com/azure/devops/pipelines/tasks/deploy/azure-resource-group-deployment
==============================================================================
```

Step 10: Now, you can see the build has been completed successfully.

Azure DevOps

: / AzureTrainings / Pipelines / Builds / AzureDevOpsTrainingProject... / #10

Search

AzureDevOpsTrainingP... +

Overview

Boards

Repos

Pipelines

Builds

Releases

Library

Task groups

Deployment groups

Test Plans

Artifacts

Project settings

#10: PLan name has been given

Release   All logs   :

Manually run today at 09:57 by V S Varma Rudra raju @ AzureTrainingWebApplicationProject |> Dev ◇ ddb959b

Logs   Summary   Tests

Agent job 1
Pool: Hosted Windows 2019 with V... · Agent: Hosted Agent

Started: 08/08/2019, 09:58:11
··· 3m 23s

Prepare job · succeeded   <1s

Initialize job · succeeded   1s

Checkout · succeeded   3s

Azure Deployment:Create Or Update Resource Group action on Training-Dev-RG · succeeded   3m 18s

Post-job: Checkout · succeeded   <1s

Finalize Job · succeeded   <1s

Report build status · succeeded   <1s

Top 30 Microsoft Azure Interview Question

A list of top 30 frequently asked **Microsoft Azure Interview Questions and answers** are given below.

1) What is Cloud Computing?

It is a platform where we can store and access our data over the internet. We can store and access our data from anywhere in the world.

2) What is Azure Cloud Service?

The Azure Cloud service offers multiple web applications in Azure; it categorizes the services and allows us the flexible scaling for our use. The Azure cloud service was launched in the year 2010. It is a dynamic cloud platform that offers development, data storage, service hosting, and service management.

3) Which service in Azure can be used to manage resources?

Azure Resource Manager is used to manage resources in Microsoft Azure. It is used to deploy, manage, and delete all the resources together using a simple JSON script.

4) What type of web application can be deployed with Azure?

Microsoft released SDKs for both Java and Ruby to allow applications written in those languages to place calls to the Azure Service Platform API to the AppFabric Service.

5) Explain Role in terms of Microsoft Azure.

Roles are nothing, but the servers are layman terms. Servers are managed, load-balanced, platform as a Service virtual machines that work together to achieve a common goal.

These roles are divided into three parts

* **Web Role:** It is used to deploy a website, using the languages supported by the IIS platform (like PHP, .NET, etc.). It was configured and customized to run web applications.
* **Worker Role:** It helps the web role to execute background processes, unlike the web Role, which is used to deploy the website.
* **VM Role:** It can be used by a user to schedule tasks and other windows services. We can use the VM role to customize the machine on which the web and worker role is running.

6) What is Virtual Machine scale sets?

It is an Azure compute resource that we can use to deploy and manage a set of identical VMs. It is easy to build large-scale services that target big compute, big data, and containerized workloads if all the VMs configured the same.

7) What are the principal segments of the Windows Azure platform?

Windows Azure has the following three principal segments:

* Windows Azure Compute: It gives a code that can be managed by the hosting environment. It provides the benefit of calculation through parts. It consists of three types of roles - Web Role, Worker Role, and VM Role.
* Windows Azure Storage: It gives four types of Storage services - Queue, Tables, Blobs, and Windows Azure Drives (VHD)
* Windows Azure AppFabric: AppFabric provides five services - Service bus, Access, Caching, Integration, and Composite.
*

8) What do you understand by autoscaling in Azure?

Azure provides the scaling of the services automatically when needed. It depends upon the use, time, and traffic that comes to our application. For example - The traffic will be higher during the examination on any exam-related application or website. Then Azure will automatically modify the setting and provide the resources as required.

9) What is the storage key?

A storage key is an authentication method that can be used to validate access for the storage service account to control data based on our prerequisites. We have an alternative to give a primary access key and a secondary access key. The main reason for using a secondary access key is to avoid downtime to the application or website.

10) Explain the SQL Azure database.

Microsoft Azure SQL database is a way to get associated with cloud services where we can store our database into the cloud. It has a similar component of SQL Server, i.e., high accessibility, versatility, and security in the core.

11) Explain cmdlet in Azure?

A cmdlet is a lightweight command that can be used as a part of the Microsoft Azure PowerShell environment. The cmdlets are summoned by the Azure PowerShell that automates the script, which is in the command line. Azure PowerShell runtime additionally invokes them automatically through Azure PowerShell APIs.

12) Explain the Migration Assistant tool in Microsoft Azure.

The migration assistant tool examines our IIS installation and recognizes the sites that can be migrated to the cloud, featuring any components which can't be relocated or not supported by the platform. This tool similarly creates websites and databases provided under the given Azure membership.

13) What is Azure SLA (Service Level Agreement)?

The SLA ensures that when we send two or more role instances for each role. Access to our cloud service will be maintained with an accuracy of 99.95 % of the time. The identification and re-correction activities will be started 99.9 % of the time whenever a role instance's procedure is not running.

14) What is Availability Set in Azure?

Availability Set is a grouping of Azure Virtual Machines. The availability set allows the Azure cloud to build and understand how the application for a user is constructed to provide availability and redundancy.

15) What steps should we take in case of drive failure?

In case of a drive failure, we should follow these steps:

❖ Unmount the drive, which allows the Azure storage object to function without fault.
❖ In the case of replacement, we will format and remount the drive.

16) Why we use VNet? Name the power states of a Virtual Machine.

We can represent our network within the cloud using VNet. VNet logically isolates our instances launched in the cloud, from the rest of our resources. The various power state of a Virtual Machine is: Running, Starting, Stopping, Deallocating, etc.

17) What is network security groups?

A network security group allows us to manage the network traffic to NIC or subnets etc. The network load will be distributed as needed if it is connected wisely.

18) What is cspack?

Cspack is a command-line tool, which is used to generate a service package file. It helps us to prepare an application for deployment, either in compute emulator or Microsoft Windows Azure.

19) Name two blobs used in Microsoft Azure.

The two types of blobs in Azure are:

- ❖ Block Blob
- ❖ Page Blob

20) Can we add an existing VM to an availability set?

No, if we want our VM to be part of an availability set, then we need to create the VM within the set.

21) How much storage can we use with a virtual machine?

In Azure, each data disk can be up to 1 TB. The number of disks we can use depends upon the size of the virtual machine. Azure Managed Disks are the recommended disk storage offering to use with Azure Virtual Machines for persistent storage of data. We can use multiple Managed Disks with each Virtual Machine.

22) How to create a VM in Azure CLI?

1. az vm create `
2. --resource-group myResourceGroup `
3. --name myVM --image win2016datacenter `
4. --admin-username Azureuser `
5. --admin-password myPassword12

23) What is Azure Search?

It is a cloud search-as-a-service solution that delegates server and infrastructure management to Microsoft, leaving us with a ready-to-use service that we can populate with our data and then use to add search to our web or mobile application. Azure search allows us to easily add a robust search experience to our applications using a simple REST API or .NET SDK without managing search infrastructure or becoming an expert in search.

24) Explain stateful and stateless micro-services for Service Fabric?

Service Fabric enables us to build applications that consist of microservices. Stateless micro-service doesn't maintain a mutable state outside a request. Azure Cloud Service's worker role is an example of a stateless service. Stateful microservice maintains a mutable, authoritative state beyond the request and its response.

25) What is a Web role in Azure Cloud Service?

A web role in Azure is a virtual machine instance running Microsoft IIS Web server that can accept and respond to HTTP or HTTPS requests.

26) Can we create a VM using Azure Resource Manager in a Virtual Network that was created using classic deployment?

It is not supported by the Azure portal. We cannot use Azure Resource Manager to deploy a Virtual machine into a virtual network which was created using classic deployment.

27) What are the options available in Azure for data storage?

Options for storing data includes:

- ❖ Azure files
- ❖ OS drive
- ❖ Scale set
- ❖ Temp drive
- ❖ Azure data service
- ❖ External data service

28) What is Azure Redis Cache?

Redis cache is an open-source, in-memory data structure store, which is used as a database, cache, and message broker. Azure Redis Cache resembles the famous open-source Redis cache. It provides access to a secure and dedicated Redis cache that is managed by Microsoft and accessible from any application inside Azure.

29) What are Redis databases?

It is a fully managed, open-source, compatible in-memory data store to power fast and scalable applications.

30) How to create a VM in PowerShell?

1. # Define a credential object
2. $cred = Get-Credential
3. # Create a virtual machine configuration

4. $vmConfig = New-AzureRmVMConfig -VMName myVM -VMSize Standard_DS2 |
5. ` Set-AzureRmVMOperatingSystem -Windows -ComputerName myVM -Credential $cred |
6. ` Set-AzureRmVMSourceImage -PublisherName MicrosoftWindowsServer -Offer WindowsServer `
7. -Skus 2016-Datacenter -Version latest | Add-AzureRmVMNetworkInterface -Id $nic.Id

Printed in Great Britain
by Amazon

41623026R00148